Bioethics, Medicine and the Criminal Law Volume 1

Who should define what constitutes ethical and lawful medical practice? Judges? Doctors? Scientists? Or someone else entirely? This volume analyses how effectively criminal law operates as a forum for resolving ethical conflict in the delivery of health care. It addresses key questions, such as: How does criminal law regulate controversial bioethical areas? What effect, positive or negative, does the use of criminal law have when regulating bioethical conflict? And can the law accommodate moral controversy? By exploring criminal law in theory and in practice and examining the broad field of bioethics as opposed to the narrower terrain of medical ethics, it offers balanced arguments that will help readers form reasoned views on the ethical legitimacy of the invocation and use of criminal law to regulate medical and scientific practice and bioethical issues.

AMEL ALGHRANI is a Lecturer in Family Law at the University of Manchester.

REBECCA BENNETT is a Senior Lecturer in Bioethics at the University of Manchester.

SUZANNE OST is Professor of Law at Lancaster University.

Cambridge Bioethics and Law

This series of books was founded by Cambridge University Press with Alexander McCall Smith as its first editor in 2003. It focuses on the law's complex and troubled relationship with medicine across both the developed and the developing world. Since the early 1990s, we have seen in many countries increasing resort to the courts by dissatisfied patients and a growing use of the courts to attempt to resolve intractable ethical dilemmas. At the same time, legislatures across the world have struggled to address the questions posed by both the successes and the failures of modern medicine, while international organisations such as WHO and UNESCO now regularly address issues of medical law.

It follows that we would expect ethical and policy questions to be integral to the analysis of the legal issues discussed in this series. The series responds to the high profile of medical law in universities, in legal and medical practice, as well as in public and political affairs. We seek to reflect the evidence that many major health-related policy debates in the UK, Europe and the international community involve a strong medical law dimension. With that in mind, we seek to address how legal analysis might have a trans-jurisdictional and international relevance. Organ retention, embryonic stem cell research, physician-assisted suicide and the allocation of resources to fund health care are but a few examples among many. The emphasis of this series is thus on matters of public concern and/or practical significance. We look for books that could make a difference to the development of medical law and enhance the role of medico-legal debate in policy circles. That is not to say that we lack interest in the important theoretical dimensions of the subject, but we aim to ensure that theoretical debate is grounded in the realities of how the law does and should interact with medicine and health care.

Books in the series

Marcus Radetzki, Marian Radetzki, Niklas Juth
Genes and Insurance: Ethical, Legal and Economic Issues

Ruth Macklin
Double Standards in Medical Research in Developing Countries

Donna Dickenson
Property in the Body: Feminist Perspectives

Matti Häyry, Ruth Chadwick, Vilhjálmur Árnason, Gardar Árnason
The Ethics and Governance of Human Genetic Databases: European Perspectives

Ken Mason
The Troubled Pregnancy: Legal Wrongs and Rights in Reproduction

Daniel Sperling
Posthumous Interests: Legal and Ethical Perspectives

Keith Syrett
Law, Legitimacy and the Rationing of Health Care

Alastair Maclean
Autonomy, Informed Consent and the Law: A Relational Change

Heather Widdows, Caroline Mullen
The Governance of Genetic Information: Who Decides?

David Price
Human Tissue in Transplantation and Research

Matti Häyry
Rationality and the Genetic Challenge: Making People Better?

Mary Donnelly
Healthcare Decision-Making and the Law: Autonomy, Capacity and the Limits of Liberalism

Anne-Maree Farrell, David Price and Muireann Quigley
Organ Shortage: Ethics, Law and Pragmatism

Sara Fovargue
Xenotransplantation and Risk: Regulating a Developing Biotechnology

John Coggon
What Makes Health Public?: A Critical Evaluation of Moral, Legal, and Political Claims in Public Health

Mark Taylor
Genetic Data and the Law: A Critical Perspective on Privacy Protection

Anne-Maree Farrell
The Politics of Blood: Ethics, Innovation and the Regulation of Risk

Stephen Smith
End-of-Life Decisions in Medical Care: Principles and Policies for Regulating the Dying Process

Michael Parker
Ethical Problems and Genetics Practice

William W. Lowrance
Privacy, Confidentiality, and Health Research

Amel Alghrani, Rebecca Bennett and Suzanne Ost
Bioethics, Medicine and the Criminal Law Volume I: The Criminal Law and Bioethical Conflict: Walking the Tightrope

Bioethics, Medicine and the Criminal Law Volume I

The Criminal Law and Bioethical Conflict: Walking the Tightrope

---

Edited by

Amel Alghrani, Rebecca Bennett and Suzanne Ost

CAMBRIDGE
UNIVERSITY PRESS

CAMBRIDGE
UNIVERSITY PRESS

University Printing House, Cambridge CB2 8BS, United Kingdom

Published in the United States of America by Cambridge University Press, New York

Cambridge University Press is part of the University of Cambridge.

It furthers the University's mission by disseminating knowledge in the pursuit of education, learning and research at the highest international levels of excellence.

www.cambridge.org
Information on this title: www.cambridge.org/9781107025127

First published 2013
Reprinted 2013

Printed in the United Kingdom by Clays, St Ives plc

A catalogue record for this publication is available from the British Library

Library of Congress Cataloguing in Publication data

Bioethics, medicine, and the criminal law : the criminal law and bioethical conflict : walking the tightrope / [edited by] Amel Alghrani, Rebecca Bennett, Suzanne Ost.
p. cm. – (Cambridge bioethics and law)
ISBN 978-1-107-02512-7 (Hardback)
1. Genetic engineering–Law and legislation–Criminal provisions.
2. Medical genetics–Law and legislation–Criminal provisions.
3. Biotechnology–Law and legislation–Criminal provisions. 4. Bioethics.
I. Alghrani, Amel. II. Bennett, Rebecca. III. Ost, Suzanne.
K3611.G46B53 2012
344.03′21–dc23

2012017169

ISBN 978-1-107-02512-7 Hardback

Contents

Contributors

AMEL ALGHRANI is a Lecturer in Family Law in the School of Law, University of Manchester. She studied law as an undergraduate and went on to qualify as a barrister in 2003, having been awarded the Yarborough Anderson Benefactors and Scholarship Award. She previously worked in the General Medical Council, but left to pursue her interest in medical law and family law. She completed a Master's degree in Healthcare Ethics & Law (University of Manchester) and a PhD under the supervision of Professors Margaret Brazier and John Harris.

DAVID ARCHARD is Professor of Philosophy, Queen's University, Belfast. He is the author of several books, and numerous chapters and articles in political, legal and applied moral philosophy. A member of the Editorial Boards of the *Journal of Applied Philosophy*, *Res Publica* and *Contemporary Political Theory*, he has also been Honorary Chair of the Society for Applied Philosophy. Since November 2005 he has been a member of the Human Fertilisation and Embryology Authority, and Deputy Chair of its Ethics and Law Advisory Committee.

REBECCA BENNETT is a Senior Lecturer at the University of Manchester. She has taught bioethics for eighteen years and has published widely on diverse bioethical issues, including antenatal HIV testing, assisted reproductive technologies, responsibility in pregnancy, the role of public consultations in ethics and law and the possibility of ethical compromise.

HAZEL BIGGS is a Professor of Law in the School of Law at the University of Southampton. She received her first degree from the University of Kent after working for several years as a radiographer and ultrasonographer in the National Health Service. Her PhD thesis examined legal and ethical aspects of euthanasia, and formed the basis of her first book, *Euthanasia, Death with Dignity and the Law* (2001).

Alongside her university activities she has been involved with NHS Research Ethics Committees since 1998. She was Chair of East Kent Local Research Ethics Committee and the Metropolitan Multi-Centre Research Ethics Committee. She is Editor-in-Chief of the *Medical Law Review.*

MARGARET BRAZIER has been a Professor of Law at the University of Manchester since 1990. She has written widely on medical law and ethics. She chaired a review of laws relating to surrogacy from 1996 to 1998 and chaired the Retained Organs Commission from 2001 to 2004. Between 2004 and June 2011, she was Editor-in-Chief of the *Medical Law Review.* She is co-author, with Suzanne Ost, of *Medicine and Bioethics in the 'Theatre' of the Criminal Process* (Cambridge University Press, 2013), and, with Emma Cave, of *Medicine, Patients and the Law* (2nd edn, 2011). She was the Principal Investigator for the AHRC-funded project, The Impact of the Criminal Process on Health Care Ethics and Practice.

SARAH CHAN is a Research Fellow in the Institute for Science, Ethics and Innovation at the University of Manchester. Her publications include (with John Harris) 'Cognitive regeneration or enhancement: the ethical issues', *Regenerative Medicine* 1 (2006), 361–6, 'Cord blood banking: what are the real issues?', *Journal of Medical Ethics* 32 (2006), 621–2, and (with John Harris) 'Evidence-based evolution: the ethics of gene therapy', *Current Opinion in Molecular Therapeutics* 8 (2006), 377–83.

JOHN COGGON is a Reader in Law at the University of Southampton. John studied for his LLB in Law at the University of Sussex, and for his PhD at Cardiff University. His research focuses on issues in legal, moral and political theory. His published research includes his book, *What Makes Health Public?* (Cambridge University Press, 2012).

SARA FOVARGUE is a Senior Lecturer in the Law School at Lancaster University. She obtained her PhD, on 'Regulating Developing Technologies: Xenotransplantation and the Law', at the University of Manchester. Her current research interests in health care law and ethics include clinical research involving human and non-human animals, developing biotechnologies, reproductive technologies, and decision-making for the 'vulnerable'. She is an Editor of the Public Policy and Law section of *Clinical Ethics* and is Assistant Editor for the *Medical Law Review.* Her recent monograph, *Xenotransplantation and Risk: Regulating a Developing Biotechnology*, is published with Cambridge University Press (2011).

MARIE FOX is a Professor of Socio-Legal Studies at Birmingham University. Her main research and teaching interests are in Health Care Law, Animal and Veterinary Law, Feminist Legal Theory, and Criminal Law. Her current research projects focus on body modification and embodiment, the legal construction of human and animal bodies and regulation of the veterinary profession. She is a coordinating editor of *Social and Legal Studies*, a member of the Advisory Board of the *Journal of Law & Society*, and a founding editor (law) of *Somatechnics*.

JOHN GRIFFITHS is a Professor of Sociology of Law (Emeritus) at the University of Groningen, the Netherlands. He studied Philosophy (University of California) and Law (Yale University) and taught Law in the United States and Ghana before joining the Faculty of Law in Groningen in 1977. His theoretical writings have been concerned with the (empirical) concept of law, legal pluralism, disputing, and the effectiveness of legal rules. In recent years, his legal and empirical research has been devoted largely to problems of the regulation of euthanasia and other socially problematic behaviour.

DAVID GURNHAM is a Reader in Law, School of Law, University of Southampton. He obtained his PhD from the University of Warwick and was previously a lecturer at the University of Manchester and the University of Reading. He teaches and researches in the areas of legal theory, health care law and ethics, and law and literature. His publications include *Memory, Imagination, Justice: Intersections of Law and Literature* (2009). He is Assistant Editor of *Contemporary Issues in Law*, and is guest editor for the *Medical Law International* (2013) twentieth anniversary special issue, 'Best Interests in an Age of Human Rights', and the *International Journal of Law in Context* (2013) issue, 'Infection in Law and Legal Discourse'.

LADY HALE was educated at Cambridge. She went on to teach at the University of Manchester, where she was appointed Professor in 1986. She also practised as a barrister for a while at the Manchester Bar. From 1984 she was a Law Commissioner, leading the Commission's work in family and mental capacity law. She was appointed a High Court judge in the Family Division in 1994, elevated to the Court of Appeal in 1999, and to the House of Lords in 2004, where she became the first ever woman to sit in the UK's highest court. She is now the first woman Justice of the Supreme Court.

JOHN HARRIS is the Lord Alliance Professor of Bioethics, and Research Director of the Institute for Science, Ethics and Innovation at the

School of Law, University of Manchester. He is the author or editor of nineteen books and over two hundred and fifty papers. He is the founder and a General Editor of a major series of books entitled Issues in Biomedical Ethics, and a new series of books, which he co-edits with John Sulston, entitled Science, Ethics and Innovation. He has published in most of the leading philosophical journals in his field and in many of the leading science journals. He is the Editor-in-Chief of the *Journal of Medical Ethics*. He frequently appears on radio and television, both in the United Kingdom and overseas, to discuss issues in biomedical ethics and medical jurisprudence.

RICHARD HUXTABLE is a Reader in Medical Ethics and Law, and Deputy Director of the Centre for Ethics in Medicine at the University of Bristol. He primarily researches in the areas of end-of-life decision-making and surgical ethics, and has published widely in medical, legal and bioethical journals, in addition to authoring the books *Euthanasia, Ethics and the Law: From Conflict to Compromise* (2007) and (with Donna Dickenson and Michael Parker) *The Cambridge Medical Ethics Workbook* (Cambridge University Press, 2nd edn, 2010). A long-standing participant in clinical ethics consultation, he is also a Trustee of the National Council for Palliative Care.

NISHAT HYDER is currently studying for a PhD in Bioethics and Medical Jurisprudence at the University of Manchester's Centre for Social Ethics and Policy. Her research concerns the legal and ethical challenges posed by artificial life technologies. Previously, she studied Law at the London School of Economics and Political Science, and went on to complete the Bar Professional Training Course at Brierley Price Prior Law School. She is particularly interested in jurisprudence, political theory, ethics and evolutionary theory, and how these disciplines relate to the emergence of new medical technologies

JOHN KEOWN holds the Rose F. Kennedy Chair in Christian Ethics at the Kennedy Institute of Ethics, Georgetown University. He graduated in Law from Cambridge and received his doctorate from Oxford. He later taught the law and ethics of medicine in the Faculty of Law at Cambridge, where he also held Fellowships at Queens' College and Churchill College. He has published widely and his work has been cited by the US Supreme Court, the English Court of Appeal and the House of Lords Select Committee on Medical Ethics. He has served on the Ethics Committee of the British Medical Association.

JOSÉ MIOLA is a Senior Lecturer in the School of Law at the University of Leicester. He joined the University of Leicester in 1998, whilst completing his doctorate at the University of Manchester. His primary research interests are in medical law and ethics. He is Assistant Editor for the *Medical Law Review*, a member of the Editorial Board of *Clinical Ethics*, and has published in a wide range of journals. He is the author of *Medical Ethics and Medical Law: A Symbiotic Relationship* (2007).

SUZANNE OST is Professor of Law in the Law School at Lancaster University. Her main research interests are in health care law and bioethics, legal and societal responses to child pornography and child sexual grooming, related aspects of criminal law and theory and law and literature. She is Editor-in-Chief of the *Medical Law Review*. She is co-author, with Margaret Brazier, of *Medicine and Bioethics in the 'Theatre' of the Criminal Process* (Cambridge University Press, 2013), and author of *Child Pornography and Sexual Grooming: Legal and Societal Responses* (Cambridge University Press, 2009). She was the Co-Investigator for the AHRC-funded project, The Impact of the Criminal Process on Health Care Ethics and Practice.

ROBERT C. SMITH is a Scottish surgeon who came to prominence in 2000 when he applied for permission to perform an amputation on a patient whose limbs were apparently healthy. Falkirk Royal Infirmary and Abbey King's Park Hospital in Stirling both refused to allow him to carry out amputations on patients he claimed to be suffering from Body Integrity Identity Disorder (BIID), a controversial psychiatric condition where sufferers experience extreme mental discomfort due to the presence of a limb they see as unwanted. Prior to the Stirling controversy, he had carried out at least two elective amputations on otherwise healthy individuals. He went on to write, with Gregg Furth (the patient whose desire for the operation had caused the furore in the first place), the first and, to date, only book uniquely about BIID, *Amputee Identity Disorder: Information, Questions, Answers and Recommendations about Self-Demand Amputation* (2000).

STEPHEN W. SMITH is a Lecturer in the Birmingham Law School. He completed his PhD at the University of Manchester on the impact of autonomy and paternalism on physician-assisted suicide and euthanasia. He previously completed a Juris Doctorate degree at Washington University School of Law in Saint Louis, Missouri, USA. He specialises in medico-ethical issues at the end of life, and has published widely on this and related issues. He is the author of *End-of-Life Decisions in Medical Care: Principles and Policies for Regulating the Dying Process* (Cambridge University Press, 2012).

Foreword

Mark Hedley
Justice of the High Court (Family Division)

Every democratic society depends for its integrity on some concept of the Rule of Law. Every democratic society learns (if only by experience) that, for the Rule of Law to be effective, two conditions must be fulfilled: first, that the substantive law must be broadly acceptable to its members; and second, that the law must be capable of enforcement. The second condition requires (at least as a last resort) the availability and implementation of criminal sanctions.

Since it is generally accepted in principle that the state has a regulatory role in the implementation and oversight of scientific (and specifically medical) research and practice, some engagement between medical ethics and the criminal law is inevitable. The real question today is not so much whether the criminal law has a role in medical research and practice or in bioethics, but rather what that role is, or should be.

There are two basic approaches to the enactment and use of the criminal law. On the one hand, legislators debate and decide on the moral criteria to be applied and in so doing define the limits of research, practice or bioethics and then enforce them by criminal penalties. Alternatively, legislators seek to permit maximum freedom within these areas (with or without a civil regulatory framework) and seek to enforce through the criminal law only evasion of that framework or the minimum standards acceptable to that society. This difference of approach may lie at the heart of many a modern controversy in this area. Whilst our society tends to follow the latter approach, it is not wholly consistent in doing so. Indeed there are many raised voices favouring the former approach, especially where debate focuses (as inevitably it must) on specific issues.

The criminal law must be recognised as something of a blunt instrument. Its substance is for Parliament, its implementation for the prosecuting authorities and its application for judges. Yet whilst appellate courts can review the law after conviction and trial judges can instruct juries in accordance with the law, the final decision in any contested case of weight lies with the jury who give no reasons for it and whose decision to acquit is effectively beyond review. Clearly this can make for real uncertainty, which itself can be compounded by the uncertainties inherent in the scientific process then under consideration.

This volume of essays correctly captures the issue in its subtitle *Walking the Tightrope*. It is a particularly difficult tightrope since, on the one hand, our society has inherent sympathy and support for medical research and practice, but, on the other hand, is likely to have firm, if often diverse, views on where the line should be drawn on specific issues. Further difficulties can be introduced in the use of justificatory concepts like 'public health' or 'human dignity', whose definitions are uncertain (or at least unagreed) and whose very provenance may sometimes be controversial. If (as I think to be so) hard cases make bad law, these are areas in which hard cases abound and therefore bad laws are an ever-present danger.

I am particularly grateful, as a judge who is not infrequently required to engage with these issues, to have been invited to write the foreword to this book, which I would wish to commend to careful study. There is no common agenda in these essays beyond a real attempt to recognise that medical research and bioethics are uneasy but probably inevitable bedfellows with the criminal law. They seek to deepen an understanding of how these bedfellows should relate in our society, making clear that this is an essential (and multidisciplinary) task that is both problematic and controversial. I found particularly helpful discussion about the place and problems of compromise in these questions, since compromise, whilst uncomfortable to the purist, is a concept innately attractive to both legislators and practising lawyers. I venture to suggest that the authors' responses to these pressing issues will put legislators, lawyers, regulators, professional leaders and all serious practitioners very much in their debt.

Acknowledgements

This edited volume is one of the major outputs of the AHRC-funded project, The Impact of the Criminal Process on Health Care Ethics and Practice, and we are extremely grateful for the Council's support.

There are a number of people who we owe a debt of gratitude to. We would especially like to thank Margot Brazier for her expert leadership as Principal Investigator on the aforementioned project, for her support and assistance regarding this edited volume, and for being such a source of knowledge, strength and friendship to us all. We would also like to thank Charles Erin for the foundational work he contributed to the volume – together with Amel, it was he who conceived of the original idea and wrote the proposal for this book. Our families and friends deserve particular thanks for allowing us the time needed to work on this project. Amel is thankful first and foremost for her son, Abrahim, who came along whilst this collection was in progress and allowed her to appreciate the finer things in life, and to her family and friends for their support. Becki would like to acknowledge the love and support given to her by Isabella, who makes everything worth it, and to thank LB for making her smile even when she thinks she can't. Suzanne is thankful for all Nick's support, and for Lily and Matthew's unwavering ability to remind her that there is much more to (and more important things in) life than her laptop. She would also like to express her gratitude to Sara Fovargue, who has been immensely supportive (both as a friend and colleague) during the duration of this project.

Finally, we are very grateful indeed to each of the contributors to this volume. It seems trite to say that without them this book would not exist. But it is true and we hope and anticipate that readers of their essays will find their contributions as engaging and thought-provoking as we have as editors.

This book is dedicated to the memory of Dr Mary Lobjoit and Professor David Price, who contributed so much to the development of medical ethics and law.

1 Introduction: When criminal law encounters bioethics: a case of tensions and incompatibilities or an apt forum for resolving ethical conflict?

Amel Alghrani, Rebecca Bennett and Suzanne Ost

Whilst health care professionals and scientists have been subject to civil law and professional regulation for some time, in England and Wales at least, questions of medical and scientific malpractice and cases involving bioethical dilemmas are increasingly coming before the *criminal* courts. This growing trend seems to have proceeded in the absence of very much by way of legal scrutiny or jurisprudential analysis.[1] Of course, some may consider the use of the criminal law as a forum to resolve bioethical dilemmas as both inevitable and wholly appropriate. Medical practice, biotechnologies and scientific research all have a major impact on our lives, and whilst they can bring and have brought us great benefits, we do not have to look too far back into the past to recall atrocities that were committed in the name of medical and scientific advancement.

The criminal law enforces moral standards and, in so doing, exemplifies the wrongs society considers especially grievous. Medical practice, biotechnologies and scientific research can involve public wrongs – that is, wrongs that should be the concern of the criminal law because they contravene defining values that the state endeavours to safeguard to ensure the good of its citizens.[2] Values that might be violated by medical practice and the development of biotechnologies include the sanctity of life and the protection of the vulnerable. Thus, arguably, the intervention

We would like to express our gratitude to Charles Erin for the part he played in laying down the foundational work for this introduction. We also thank Margot Brazier for all her valued help and support whilst putting this collection together.

1 The way in which the criminal law regulates health care practice has been explored in C. A. Erin and S. Ost (eds.), *The Criminal Justice System and Health Care* (Oxford University Press, 2007), a volume that focuses more on health care in general, as opposed to bioethics.

2 See S. E. Marshall and R. A. Duff, 'Criminalization and Shared Wrongs' (1998) XI(1) *Canadian Journal of Law and Jurisprudence*, 7, 20.

of the criminal law is not only apposite, it is *required* and there is an inevitability about its intervention in any society that utilises the criminal law to reflect its basic values and punish those who culpably cause others harm or a risk of harm.

Others might contend that resort to the criminal process in these areas constitutes morally supportable redress to what they see as a prior conspiracy of professions, whereby the law almost 'naturally' deferred to the perceived authority of those who practise in the rarefied environment of medicine.[3] And whilst some might not wish to make medical professionals who commit an error accountable to the criminal law,[4] as Oliver Quick has recently observed: 'Not even the staunchest sympathiser of the medical profession would call for a blanket ban on criminalisation. Some events will always be beyond the pale and warrant a criminal response.'[5]

A further, more sceptical view may be that the criminal law is being employed as part of the state's exercise of biopower,[6] to regulate and subjugate our bodies by placing clear limits on what can and cannot be done, regardless of our consent. Whether or not such a Foucauldian argument convinces, it seems only right that the growing tendency to involve the criminal law should be questioned, and this collection of essays by leading legal and philosophical thinkers sets out to do exactly that.

Given the rich philosophical underpinnings of some criminal law and bioethical theories, and the common appearance of utilitarian and deontological approaches and liberalist ideals in both, it would seem that, at least on the face of it, parallels can be found between the two.[7] Moreover, bioethical[8] and criminal law theory both offer fertile grounds for analysis of the law's social role. For example, criminal law theorists have explored the matters of whether the criminal law deters and

[3] I. Kennedy, *The Unmasking of Medicine* (London: Allen and Unwin, 1981).

[4] See J. K. Mason and G. T. Laurie, *Mason and McCall Smith's Law and Medical Ethics* (8th edn) (Oxford University Press, 2010), p. 167. See also Archard in this volume.

[5] O. Quick, 'Medicine, Mistakes and Manslaughter: A Criminal Combination?' (2010) 69(1) *Cambridge Law Quarterly*, 186.

[6] M. Foucault, *The History of Sexuality Vol. 1: The Will to Knowledge* (New York: Pantheon, 1978).

[7] M. Brazier and S. Ost, *Medicine and Bioethics in the 'Theatre' of the Criminal Process* (Cambridge University Press, in press). See especially Part 3.

[8] When we talk about 'bioethical theory' or 'bioethicists' in this introductory chapter, we are referring to what might be termed 'philosophical bioethical theory' and 'philosophical bioethicists'. Philosophical bioethics uses the methodology of moral philosophy to analyse ethical issues that arise in the biosciences, including medicine and health care. There is, of course, a more interdisciplinary use of the word 'bioethical' or 'bioethicist' that does not necessarily hold the methodology of moral philosophy as its primary focus.

whether its purpose should be to deal out retribution and to punish, or to achieve restorative justice.[9] Bioethicists have considered the law's role in resolving difficult cases involving ethical conflict.[10] And there is much to be found in criminal law theory that is of direct relevance to the kind of debates present in bioethics (for example, the sanctity of life[11]) and vice versa (autonomy,[12] for instance). But debate in each usually remains confined within the boundaries of the discipline, aimed at an audience either of criminal lawyers *or* of bioethicists. In order to offer a richer assessment of the appropriateness of the criminal law's regulation of medical practice and bioethical issues, there is a need for the two disciplines to talk to each other.

This volume offers balanced arguments that will help the reader form a reasoned view on the ethical legitimacy of the invocation and use of the criminal law to regulate medical practice and issues of bioethics. It aims to shed light on the question of who *should* define what constitutes ethical, and thus lawful, medical practice – judges, the legislature, doctors, scientists or someone else? To this end, it analyses how effectively the criminal justice system can and does operate as a forum for resolving ethical conflict in the delivery of health care. Key questions that are addressed include: How does the criminal law regulate controversial bioethical areas? Is the use of the criminal law in these areas appropriate or desirable? What effect, positive or negative, does the use of the criminal law have when regulating bioethical conflict? Can the law accommodate moral controversy? Does compromise offer a way forward? The volume explores criminal law in theory and in practice, and the broad field of bioethics rather than the narrower terrain of medical ethics. All the essays are interrelated by their analysis of the criminal law's regulation of bioethical areas; however, whilst numerous essays focus on criminal law within the specific context of health care, others address scientific research and biotechnologies. The book is divided into the following four parts, framed around what we conceive to be some of the most dominant issues involved when criminal law and bioethics encounter each other.

[9] See, e.g., A. von Hirsch, A. E. Bottoms, E. Burney and P.-O. Wikström, *Criminal Deterrence and Sentence Severity* (Oxford: Hart, 2000); D. Wood, 'Retribution, Crime Reduction and the Justification of Punishment' (2002) 22 *Oxford Journal of Legal Studies*, 301; M. Munn, 'Restorative Justice: An Alternative to Vengeance' (1993) 20 *American Journal of Criminal Law*, 99.

[10] See M. Powers, 'Bioethics as Politics' (2005) 15 *Kennedy Institute of Ethics Journal*, 305, 306.

[11] See, e.g., G. Williams, *The Sanctity of Life and the Criminal Law* (London: Faber and Faber, 1958).

[12] See, e.g., O. O'Neill, *Autonomy and Trust in Bioethics* (Cambridge University Press, 2002).

Part I: Death, dying and the criminal law

Medically assisted dying is perhaps one of the most obvious examples of a phenomenon that causes the fields of bioethics and criminal law to intersect. It is thus, ironically, the ending of life that marks our starting point. For all the protracted, intense dispute engendered by the involvement of the medical profession in the premeditated ending of a person's life, about the only area of agreement between those ranged on all sides seems to lie in the acceptance that this is a very sensitive area of social and legal policy, and we need safeguards in law to protect the vulnerable. Although the debate is far from static, the basic lines of argument have not changed much in decades: principles of personal autonomy and of sanctity of life still dominate. Is the current legal position truly reflective of current attitudes of the general public, the medical profession or the judiciary? By whom should our lawmakers be guided? Does criminal liability of doctors in this area make for 'good medicine'? How successfully does the criminal law accommodate the moral controversy that surrounds medically assisted dying? Whilst it may seem that the only way to deal with this is to go one way or the other – that is, to criminalise or decriminalise – there is also a subtler middle-ground approach. Each of these approaches is advocated by the three contributors in this part. The decriminalisation and legalisation of euthanasia are John Griffiths' recommendations. Utilising the Netherlands as a case study, Griffiths contends that the criminal law's involvement in doctors' death-causing behaviour is dangerous. Furthermore, because some types of this behaviour are lawful and there are no moral, legal or public policy differences between these and euthanasia, legalisation of the latter should be the way forward. What he highlights as the most difficult matter to decide is ascertaining what should constitute 'appropriate' cases in which euthanasia is justifiable.

In marked contrast to Griffiths, John Keown sees the intervention of the criminal law as wholly appropriate and necessary. For him, the only advisable way to go is to ensure the continued criminalisation of euthanasia, a position he explains through challenging five commonly raised arguments in favour of decriminalisation. Keown sees the prohibition on intentional killing as core to law and society, and argues that its reflection in the criminal law's response to euthanasia is both right and proper and means that no conflict exists with bioethics.

Aiming to move the assisted dying debate forward, Richard Huxtable paves a middle way between the polarised approaches reflected in the essays by Griffiths and Keown. Whilst the idea of compromising is likely to face hostility from those who stand on both sides of the assisted dying debate, given their strongly held convictions, Huxtable defends his view that compromise is the way forward in the face of conflicting opinion in the bioethical literature. The compromise he proposes involves basing a new offence of mercy killing on excuse as opposed to justification, since he contends that the notion of excuse offers a means to embrace the ambivalence that lies at the heart of the debate and the criminal law's response to assisted dying. Huxtable's ship of compromise traverses the rough waters between prohibition and permission, demanding concessions from both opponents and supporters of assisted dying, in order to achieve a new, 'morally sensitive' approach to this phenomenon in criminal law and bioethics.

Part II: Freedom and autonomy: when consent is not enough

The principle of respect for individual autonomy is a core ethical principle of modern medical ethics and bioethics more generally, and this is reflected clearly in medical law. We no longer live in a world where either the doctor or the state is thought to know best when it comes to our individual choices about our lives. Thus, it is normally assumed (both in law and ethics) that if an individual has sufficient mental capacity to understand his decision and its consequences, is not unduly influenced, has sufficient information to make a considered choice and is unlikely to cause serious harm to others, he should be allowed the freedom to make such a choice about his life, however foolish or risky we may think that choice is. This freedom to make such decisions and retain control over our lives is, in general, viewed as more important than protection from the possible risky consequences of these choices. Thus, in the medical arena, the law supports a competent patient's decision regarding whether to accept or reject medical treatment, regardless of any possible difficult or dangerous consequences of such a choice.[13]

Part II considers whether criminal law's often more paternalistic approach to acts that constitute occasioning actual and grievous bodily harm can be reconciled with the trend in both the bioethical literature and medical law to view respect for individual autonomy as the paramount consideration. It asks whether we should we accept that, in such situations, medical and legal paternalism is justified, that personal autonomy and informed consent are not conceptually peremptory.

In the first essay in this part, Surgeon Robert Smith considers the ethical and legal aspects of his involvement with patients with Body

[13] See cases such as *B* v. *An NHS Hospital Trust* [2002] EWHC 429 (Fam).

Integrity Identity Disorder (BIID), the desire for amputation of a healthy limb. He argues that a society that holds respect for individual autonomy as a core ethical and legal principle should allow individuals with BIID to undergo amputation, however foolish this decision may seem to others. Smith contends that patients with BIID are generally well-informed, appear to have the required mental capacity to make medical decisions and are fully aware of the consequences of their wish to amputate a limb. Further, he claims that whilst limb amputation may seem to others to be a great harm, to patients with BIID, limb amputation can be a great personal benefit and will not necessarily cost society in terms of resources, resulting in happier, more productive members of society. Currently, although unprecedented, criminal sanctions could be taken against a surgeon providing this kind of amputation. Smith argues that such a use of the criminal law is not an appropriate response to BIID and its treatment. He suggests that BIID should be recognised as a disorder treatable by surgery in a similar way that Gender Identity Disorder is, rather than a cause for the involvement of the criminal law.

Cases of sexual transmission of diseases such as HIV raise interesting and difficult issues when it comes to respect for individual autonomy. Whilst we normally assume that individuals' decisions should be respected, where these decisions put others at risk of serious harm there may well be some justification for legal and even criminal sanctions. In 'Risky sex and "manly diversions": contours of consent in HIV transmission and rough horseplay cases', David Gurnham notes how inconsistent the legal approaches taken by the criminal courts are in deciding whether or not consent is enough when dealing with the infliction of grievous bodily harm during sexual encounters. He highlights how, depending on the types of behaviour and the context, the courts often deem bodily harm inflicted unintentionally through 'rough and undisciplined horseplay' lawful, even when bodily harm is foreseen and consent wrongly presumed; yet unsafe sexual activity where one party is aware s/he is infected with HIV and recklessly transmits it to the other party is deemed unlawful, unless 'informed consent' was given to the risk. In this context, it is clear that legal paternalism will prevail over individual autonomy, and even in the context of private sexual encounters, the criminal common law holds that there are indeed certain harms where consent simply is not enough.

Whilst it is clear that the criminal law's involvement deems what sexual conduct is lawful or unlawful when grievous bodily harm is inflicted, should the role of the criminal law be extended to determine whether other sorts of sexual conduct ought to be unlawful? Take, for instance, sexual activity between individuals where there is an imbalance of power that might be regarded as exploitative and could negate consent – one example being the relationship between doctor and patient. Considering whether criminal law should be extended further, Suzanne Ost and Hazel Biggs explore the doctor–patient relationship and the controversial issue of maintaining sexual boundaries. The ethical and regulatory guidance on maintaining sexual boundaries between health care professionals and their patients takes the position that *any* engagement in sexual activity with a patient by a doctor is inappropriate and damaging. They explore why this might be the case and whether consent can ever really freely be given in the context of this relationship. Should a doctor who enters into a consensual sexual relationship with a patient incur the wrath of the criminal law due to the inherent imbalance of power between the two? Their conclusion is that the criminal law's intervention is warranted in such cases when the doctor obtains the patient's consent through deception, inducement or coercion, and they highlight the possible limitations of the criminal law in England and Wales in this regard.

Part III: Criminalising biomedical science

Law and bioethics clearly come together in the context of scientific endeavour, and controversial applications of science continue to raise challenging questions for both legislators and philosophers. Advances in the life sciences proceed apace in the early twenty-first century, and it is hard to ignore current and potential contributions to medical practice. Yet the relationship between science and society is at times an uneasy one. Concerns and controversy over scientific research and new technologies and advances often lead to calls for stronger regulation. Part III of the volume considers the foundations for the regulation of scientific research, and the suitability of the criminal courts for this role. As Alghrani and Chan note in their essay, '"Scientists in the dock": regulating science', negotiating the legitimate bounds of scientific activity requires a precarious balancing act, for whilst on the one hand there is the desire to encourage innovative research, on the other there is the need to minimise harms that research might cause, and ensure accountability so as to allay public concerns. In their exploration of the regulation of science, they focus on research in the controversial areas of human reproduction and human tissue, examining the use of criminal law to regulate these areas and whether it is an appropriate or effective regulatory mechanism for science. In these areas, they argue the criminal law is used less as a direct regulatory mechanism and more as a means of gaining public trust and legitimising science that would, if not subjected to the powerful symbolic force of criminal regulation, arouse even greater public concern.

The matter of balancing risk and benefit is also tackled by Sara Fovargue. Her essay presents xenotransplantation as a site of bioethical conflict because, whilst successful xenotransplants would benefit recipients and society, there are real risks. Given this, whilst recognising the dangers of overcriminalisation, she conceives criminal law regulation as imperative to protect public health. Indeed, Fovargue goes so far as to contend that without the force of the criminal law to safeguard against the threat to public health, genetically engineered solid organ clinical xenotransplants should not be permitted to occur.

Whilst the bioethical conflict generated by xenotransplantation relates to the development of biotechnologies that might save lives but also threaten health, there are of course other scientific developments that may not pose such a significant danger to health, but still raise controversy, such as the emerging use of drugs that can enhance individual performance/cognitive ability. Such developments are not considered by some to be in any way the business of the criminal law. Nishat Hyder and John Harris certainly advocate such a view in their essay, 'The criminal law and enhancement: none of the law's business?' They argue that chemical cognitive enhancement is none of the law's business for two main reasons: the first is that it is both a right and a duty of the individual to enhance oneself in a wide variety of ways; second, they argue that the availability of chemical cognitive enhancements promotes the societal interest in a more productive and efficient workforce, less affected by stress and ill health. Thus, they argue that enhancements are in both the personal and the public interest and should be the prudential concern of individuals as well as governments. The criminal law can only thwart the public interest in personal aspiration by interference.

On a different tack, Stephen Smith examines the use of the concept of dignity by the criminal law to regulate scientific research. Smith argues that interference with dignity may constitute a harm, and thus the concept of human dignity may be invoked as an 'ethical brake' to provide some protection to embryos, those in persistent vegetative states, infants and individuals with advanced dementia – that is, where respect for individual autonomy cannot be used. However, whilst this concept of dignity may be seen as a useful one, it is notoriously difficult to pin down. The purpose of Smith's chapter is to clarify the concept of dignity, particularly exploring what entities have human dignity and why, so that we can then use this concept of dignity in a meaningful way. Smith concludes that as there is nothing that humans inherently share that would provide the grounding for a concept of human dignity, we should base the concept on a socially constructed value, so that the harm that may be avoided by the concept of dignity is the harm of being unjustifiably treated as outside of the human community.

Part IV: Bioethics and criminal law in the dock

The final part of the volume examines the suitability of the criminal law to regulate ethical matters and medical practice in terms of moral and jurisprudential bases. Contributors examine how the attitudes of the judiciary have changed over time on the role of the criminal law in matters of bioethical conflict, their understanding of contemporary bioethics and the difficulties judges face in grappling with moral controversy. They consider the suitability of the criminal courts for the role of adjudicating bioethical conflict and whether the criminal law's approach to regulation of bioethical controversy is potentially damaging to society.

Medical law raises many ethical issues that often raise moral questions that are difficult for judges to grapple with. As Mason and Laurie say, 'Medical law is catalysed by moral issues'.[14] Whilst moral issues are present in many of the non-bioethical disputes judges are asked to determine, Margaret Brazier examines why it may be that the moral questions that medicine and science provoke are especially tricky. She examines whether the common law can accommodate such moral controversy principally through the prism of abortion laws, which explicitly cast abortion as criminal by way of the Offences Against the Person Act 1861, whilst offering defences to doctors under the Abortion Act 1967 (as amended). She notes the limitations of the law, and indeed science, in providing answers to some of society's most challenging questions, such as on the sanctity of life, and argues that when dealing with issues to which there are no 'right' answers, the challenge for the law is to pay equal consideration to diverging views and, in so doing, resolve moral controversy in a way that best fits the society in which it exists.

Across the water in Northern Ireland, the Abortion Act 1967 and defences therein do not apply – in that jurisdiction there are thus no statutory grounds for abortion, and termination of pregnancy is a criminal offence subject only to minor exceptions carved out by the common law. Use of the criminal law to govern this issue continues to create enormous injustices for women residing there. Engaging with the politics of abortion reform in this jurisdiction, Marie Fox argues that feminist arguments should focus on the inappropriateness of deploying the criminal law to

14 Mason and Laurie, *Mason and McCall Smith's Law*, p. 2.

regulate decisions about pregnancy, by highlighting the material realities of criminalising abortion. Second, she cautions that reforms that bypass the moral dimensions of the debate and are couched in consumerist rhetoric about choice and services may prove counterproductive in the Northern Irish context. Rather, she contends that legal reforms must be framed with historical and cultural contexts in mind, and suggests that framing pro-abortion arguments as a matter of discrimination and injustice – discourses that have a particular resonance and appeal in Northern Ireland – may be more likely to command support.

Individual doctors working at the frontiers of medical ethics and practice may also incur the wrath of the criminal law. A doctor who commits a fatal error that is judged to be grossly negligent may find himself in the dock charged with gross negligence manslaughter. When this occurs, are doctors 'special', or, indeed, should they be regarded as so for the purposes of the criminal law because of the nature of the work undertaken? In 'The impact of the loss of deference towards the medical profession', José Miola considers the courts' deference towards the medical profession in general terms. He argues that in a shift from the deferential approach in the past, the civil courts have deliberately sought to wrestle control back from medical practitioners and this has also filtered through to the criminal sphere, as evidenced by the rise in number of gross negligence cases. Nevertheless, doctors are still treated no worse (and arguably better) than other professionals. Whilst he contends that in this regard the law is moving in the right direction, he notes that this does not mean that criminal law is the correct forum for addressing medical errors – particularly those involving negligence rather than intentional harm.

Offering a philosopher's analysis of the issue of whether grossly negligent doctors who cause their patients' deaths merit special treatment, David Archard places emphasis on the characteristics of such behaviour, alongside the particular context of the medical professional's role and the value of this role. In his view, whilst 'malicious doctoring' should rightly be prosecuted by the criminal law, we should think carefully about whether the prosecution of negligent doctors leads to the fair distribution of burdens and benefits in the criminal justice system.

Public health is the particular focus for John Coggon in 'All to the good? Criminality, politics, and public health'. He explores the jurisdiction of the law and the state over health, and considers whether public health measures should include the utilisation of the criminal law in a political liberal state. Whilst recognising a logical and defensible role for the criminal law, Coggon's argument echoes the liberal paradigm; criminal law should not be used lightly, especially in the case of offences pertaining to health that generate strong moral condemnation, and in the case of self-regarding 'unhealthy' behaviour. He cautions against the assumption (without careful analysis of the issue) that the protection of health is a sound justification for criminal law regulation.

Bringing the book to a close, Lady Hale's essay draws attention to the fact that, as with bioethics, law is not without its own controversy, and judges themselves often have to walk the tightrope: difficult conundrums posed by human rights and the task of interpreting the law lead judges into moral and political controversy. Lady Hale reminds us that it is not for judges to decide what behaviour should be criminal, and their subjective views about a defendant's moral culpability can have no bearing on a case. Judges do not court moral controversy, but the challenges posed by cases such as *Pretty*[15] and *Purdy*[16] mean that they cannot escape it.

Themes and connections

Finally here, we draw attention to the themes that connect the essays in this work. Virtually all of the contributors explore the theme of what the appropriate limits of the criminal law should be (e.g., Archard, Coggon, Fovargue, Griffiths, Gurnham, Hyder/Harris, Huxtable, Keown, Ost/Biggs and R. Smith). A number of essays explore whether the role/character of the actor makes a difference to their retrospective/outcome responsibility[17] – that is, when we consider it appropriate to blame them, and, indeed, hold them criminally liable – for their actions (e.g., Archard, Miola and Ost/Biggs). The theme of moral controversy connects other essays, such as those by Brazier, Hale, Hyder/Harris and R. Smith. Paternalism on the part of the state in regulating areas of ethical controversy is also a common connector, as is the extent to which it should override individual autonomy (e.g., R. Smith and Gurnham) and the extent to which, and on what grounds, certain scientific research/endeavours should be permissible or not (Alghrani/Chan, Fovargue and S. Smith).

Academic consideration of the issues raised by the criminal law's intervention with bioethics is still in its infancy, and, consequently, what this volume cannot provide is all the answers to the question posed in

[15] *R (Pretty)* v. *Director of Public Prosecutions* [2001] UKHL 61.

[16] *R (Purdy)* v. *Director of Public Prosecutions* [2009] UKHL 45.

[17] G. Williams, 'Responsibility', Internet Encyclopaedia of Philosophy (2006, last updated 2010; see www.iep.utm.edu/responsi-3); N. Vincent, 'A Structured Taxonomy of Responsibility Concepts', in N. Vincent, I. van der Poel and J. Van der Hoven (eds.), *Moral Responsibility: Beyond Free Will and Determinism* (London: Springer, 2011), p. 17.

the title of this introduction. What we hope, however, is that this volume marks the first part of an important journey that will engage and connect criminal law theorists and bioethicists. Moreover, it offers suggested ways forward to resolve tensions encountered when criminal law meets bioethics.

Part I

Death, dying and the criminal law

2 Euthanasia and assisted suicide should, when properly performed by a doctor in an appropriate case, be decriminalised

John Griffiths

Introduction

I begin this essay with some introductory remarks. The first concerns the wider medical context of different sorts of death-causing medical behaviour by doctors, of which euthanasia and assisted suicide are but a very small part. Then I discuss the difference between 'decriminalisation' and 'legalisation'. Next I give sketches of the deathbed situations where such medical behaviour takes place, and the dangers associated with them.

After these preliminaries I turn to the thesis under debate. I shall defend three specific propositions: (1) the argument for *decriminalisation* is that death-causing behaviour by doctors is too dangerous to be left to the criminal law; (2) the argument for *legalisation* is that there are no significant moral, legal or public policy differences between the various ways a doctor can cause the death of a patient; (3) the most difficult aspect of a legalised regime is the limitation to 'an appropriate case'. In presenting the argument, I shall rely primarily on data from the Netherlands.

The medical context

To debate the legal status of euthanasia and physician-assisted suicide fruitfully, we need to have clearly in mind the medical context in which such behaviour would take place. The time and manner of roughly half of all deaths are intentionally determined by something the patient's doctor does. We can call the medical behaviour involved 'medical behaviour that potentially shortens life' (MBPSL). There are a number of sorts of MBPSL, depending on the circumstances and on exactly what the doctor does. Their frequencies vary considerably over time and

This and the following chapter were written after an oral debate between the two authors at the AHRC-funded conference, *Good, Bad or Indifferent? Medicine and the Criminal Process*, at the University of Manchester on 4 November 2009. However, neither has read nor reacted to the other's contribution to this volume.

Table 2.1. *Sorts and frequencies of MBPSL*

Sorts	Frequency (percentages of all deaths; data derived from doctors)	
	NL 2005	UK 2004
Termination of life on request:		
Euthanasia	1.7	0.16
Physician-assisted suicide	0.1	0.00
Termination of life without request	0.4	0.33
Pain relief with life-shortening effect	25	33
Abstention from life-prolonging treatment	16	30
Total MBPSL	43	64

Source: J. Griffiths, H. Weyers and M. Adams, *Euthanasia and Law in Europe* (Oxford: Hart, 2008) pp. 154, 489.

between countries. Table 2.1 provides a quick overview of the various sorts, with the most recent estimates of the frequency of each sort in the Netherlands and the UK.

In the rest of this essay, I shall generally refer to both euthanasia and physician-assisted suicide as 'euthanasia'. In my view, the differences are important,[1] but for the purposes of the present debate, not particularly relevant. As Table 2.1 shows, euthanasia is a relatively unimportant part of the whole of MBPSL in both countries, although according to this sort of data – which are the best we have – more common in the Netherlands, where it is legal, than in the UK, where it is not. On the other hand, all MBPSL taken together are responsible for many more deaths in the UK than in the Netherlands, because of the higher frequencies in the UK of death due to pain relief and to abstention.

'Decriminalisation' and 'legalisation'

A second preliminary remark is that although the terms are often loosely used as if they were interchangeable, 'decriminalisation' is not the same thing as 'legalisation'. Each is possible with or without the other.

'Decriminalisation' refers to the specific type of control to which behaviour is subjected and leaves open the question of whether the behaviour is legal or not. Thus, decriminalisation of euthanasia could

[1] See J. Griffiths, 'Physician-Assisted Suicide in the Netherlands and Belgium', in D. Birnbacher and E. Dahl (eds.), *Giving Death a Helping Hand: Physician-Assisted Suicide and Public Policy, an International Perspective* (Dordrecht: Springer, 2008), p. 77.

perfectly well be accompanied by a prohibition enforced in some other way than with the criminal law. In Switzerland, for example, assisted suicide is not a criminal offence, but direct participation by a doctor is prohibited by the code of medical ethics. This is the reason the Swiss practice is in the hands of laypersons.[2]

'Legalisation', on the other hand, leaves open the question of how the rules governing the legal practice are to be enforced. In other words, 'legalisation' does not necessarily entail 'decriminalisation': the rules that apply to legalised euthanasia could perfectly well be enforced with criminal sanctions. This was in fact the situation in the Netherlands, after euthanasia had effectively been legalised in 1984 by the judicial recognition of a justification. The 'rules of careful practice' that defined the scope of the justification remained subject to criminal enforcement until, in 2002, legislation largely decriminalised their enforcement and gave primary responsibility for control to Regional Review Committees.[3]

My personal view is that both decriminalisation and legalisation can be desirable, but they are two different things. The arguments for one or the other are rather different, and it can be desirable to do either one of them even if one does not do the other. Legalisation of euthanasia is desirable even if enforcement of the rules to which it is subject remains a criminal matter, and decriminalisation is desirable even if euthanasia remains illegal.

Three scenarios of dying and their dangers

Let me now make concrete the issue being debated. I shall do this by contrasting ways one might die as a result of something one's doctor does or does not do. Put yourself in the position of a very sick patient. You know you are dying, let us suppose of cancer, and you are suffering from a variety of symptoms that cause both pain and other sorts of suffering. You may have a variety of reasons for wanting to die in a particular way, but I would like to focus on the question of *personal safety*. Which of the following scenarios would make you feel more comfortable?

Scenario 1

Like most people, you have made no preparations for the situation you find yourself in. Because you are close to death, and as a result of

[2] See G. Bosshard, 'Switzerland', in J. Griffiths, H. Weyers and M. Adams, *Euthanasia and Law in Europe* (Oxford: Hart, 2008), Chapter 16.

[3] For the history of legalisation and decriminalisation in the Netherlands, see Griffiths *et al.*, *Euthanasia and Law*, and J. Griffiths, A. Bood and H. Weyers, *Euthanasia and Law in the Netherlands* (Amsterdam University Press, 1998).

the pain medication you are getting, it is not easy for your doctor to communicate with you. You are only dimly or intermittently aware of what is going on.

There may be a number of reasons why it would be humane or convenient that you die sooner rather than later: you seem to be suffering seriously; your loved ones, gathered around your bedside, are exhausted and cannot take much more; there is a shortage of beds and nursing staff; your doctor has other patients to attend to. In the United States, we would (perhaps until very recently) have had to add to this list: your insurance coverage has run out.

Neither the medical profession nor the hospital has any kind of protocol regarding what to do in your situation. With or without consulting you or your loved ones, your doctor decides to increase your pain medication drastically, expecting you to die earlier than you would otherwise have done. Or he decides to withdraw some form of life-sustaining treatment. Either way, you die shortly thereafter. Your death is reported as a 'natural' one and no one checks to see that the decision-making procedure and the decisions made were appropriate.

According to the available quantitative data that we saw in Table 2.1, almost two-thirds of all deaths in the UK and about two-fifths of all deaths in the Netherlands nowadays take place in roughly this way.

Scenario 2

In the second scenario, you have prepared yourself in advance for the situation you now find yourself in – perhaps years before you became ill. You may have made a written request for euthanasia in such a situation and you will have discussed it with your doctor far in advance. Whether or not you made advance arrangements, while still able to do so you make a concrete request for euthanasia. Your doctor considers whether in the circumstances your situation fits the legal criteria. Having decided that this is the case, he requests a second, independent doctor to examine you and to make a judgement. The second doctor agrees that your situation meets the requirements. At the appointed time, with your loved ones present, your doctor gives you a sedative and then a lethal injection. You die a few minutes later. Your doctor makes a full report of your death to a review committee that is independent from the criminal law authorities. The committee ascertains that the legal requirements were met. (If they were not, it refers the case to the prosecutors.)

This is a pretty fair picture of how people in the Netherlands (and Belgium) currently die from euthanasia. About 2 per cent of all deaths in the Netherlands take place in this way. While reality is, of course, not always so nice and tidy, the shadow-side of Dutch euthanasia practice is largely to be found where an outsider might not expect it: in two-thirds of all cases of a concrete request from the patient the doctor refuses or fails to carry it out.[4]

Scenario 3

There is a third possible scenario that may or may not apply to your situation. Some time in advance you have given written instructions concerning the treatments you do not want if you are in such a situation. If you were well-advised, you have also appointed a representative to see that your wishes are carried out. If you are still capable of communication, you can also orally refuse treatment on the spot. Treatment is withdrawn. You die some time thereafter.

Apart from the choice of method for accomplishing your wish to die, this scenario looks rather like euthanasia, except that your refusal of treatment is in principle legally binding on your doctor, and there will be no external review of what took place. It seems from the Dutch data that a request by the patient is involved in rather less than a quarter of all cases of death due to withdrawing or withholding treatment,[5] which means that about 4 per cent of all deaths are due to refusal of life-prolonging treatment by the patient.

In light of the above three scenarios, the argument I am going to make is simple: of all deaths at the hands of one's doctor, the vast majority look like the first scenario and take place in a way that is highly unsafe and cries out for regulation and enforcement. The only important exception to this is euthanasia as practised in the Netherlands and Belgium. Unlike most deaths due to pain relief or abstention, the practice is a reasonably safe one. This is because it has been legalised and control has been decriminalised.

[4] Unfortunately, the Dutch data (see Griffiths *et al.*, *Euthanasia and Law*, pp. 170–1) do not permit a quantitative estimate of the number of cases of refusal in which the reason is not the patient's failure to 'qualify', but unwillingness on the part of the doctor. Qualitative data suggest that reluctance among doctors is widespread, and that this manifests itself both in straightforward refusals (which leave to the patient the possibility – unfortunately often only theoretical – of transfer to another doctor) and in the form of avoidance, disingenuousness, procrastination and so forth (see Griffiths *et al.*, *Euthanasia and Law in the Netherlands*, pp. 246–8). For some recent quantitative indications (in the case of people who take matters into their own hands and carry out suicide with the support of family and friends), see Griffiths *et al.*, *Euthanasia and Law*, at pp. 183, 188.

[5] See Griffiths *et al.*, *Euthanasia and Law*, p. 162 (data from 2001).

The argument for decriminalisation

It is commonly supposed that if a value – in our case, human life – is really important, only the criminal law is sufficient for its protection. I shall argue, on the contrary, that the criminal law is an ineffective way to regulate intentional death-causing behaviour by doctors. The argument for decriminalisation (with or without legalisation) is that intentional death-causing behaviour by doctors is too dangerous to be left to the criminal law.

I am not going to argue the point that behaviour of doctors that intentionally causes the death of patients is a socially precarious matter. The possibility of practices that could not stand the light of day is so immanent in such behaviour that some form of societal control is essential. I take this to be obvious.

If the patient asked for (or is said to have asked for) what was done, this does change the matter somewhat, but substantial danger of abuse remains. This judgment seems to me to apply equally to all MBPSL. I cannot see that the fact that some of the resulting deaths are attributed to 'normal medical practice' and reported as 'natural' makes any difference whatsoever as far as public safety is concerned.

Granted the need for regulation, my argument is that the criminal law is an expensive and cumbersome instrument, an ineffective way to regulate this sort of practice, and one that causes more harm than it prevents. Let me quickly note the most important considerations:

- The review capacity of the criminal law is not even remotely adequate to the task. Enforcement of the rules concerning MBPSL would require some attention to every such case. Before decriminalisation of euthanasia in the Netherlands, the prosecutorial authorities were not capable of seriously reviewing the roughly 2,000 cases being reported each year.[6] To deal with all MBPSL would entail reviewing almost 60,000 cases per year in the Netherlands, and well over 300,000 per year in England and Wales.[7] Such numbers mean that even cursory examination of each case would totally swamp the criminal law institutions.

[6] See Griffiths *et al.*, *Euthanasia and Law*, p. 207 (in 1997, about 2 per cent of all reported cases were fully reviewed). I leave aside here the additional objection that such review was to a significant extent incompetent, due to an absence of medical expertise among the reviewers.

[7] See Griffiths *et al.*, *Euthanasia and Law*, p. 154; UK Mortality Statistics 2005, National Statistics Series DH2 No. 32 (London: Office for National Statistics, 2006) (total deaths × 0.64; see Table 2.1).

- The criminal law institutions have no way of identifying cases of medical end-of-life practice that require attention. The criminal law is in general reactive: it deals with cases brought to its attention, which is seldom the case of behaviour in private. One only has to think of the invisibility over many years of Dr Shipman's murderous practices[8] to realise how feeble an instrument the criminal law is for dealing with this sort of thing.
- Even when cases of illegal medical practice come to the attention of the criminal law authorities, their respect for the professional status of doctors and their awareness of how drastic the sanctions available to them would be if actually applied, give prosecutors and judges serious reservations about the use of the criminal law except in the most egregious cases. (I suggest that potential sources of complaints or information – nurses, for example – feel the same constraints.)
- There are some technical aspects of the criminal law that make it unsuited to the situation where the death of a dying person is brought about by a doctor. Perhaps the most serious is the requirement of proving causation. If taken seriously, proving that it is what the doctor did – rather than the patient's underlying condition – that caused the patient's death will in many cases be impossible.
- Given its lack of capacity to do much good, the fact that the criminal law can do serious damage to patient safety is a matter of special concern. Serious investigators like Magnusson[9] have shown how the criminalisation of euthanasia fosters a secret, underground practice that produces just the dangers we most want to prevent: underground euthanasia is often medically sloppy and ethically dubious, and any kind of control is non-existent.

These considerations lead to the conclusion that patients who die from euthanasia in the hands of their doctors in the Netherlands, where societal control of euthanasia is largely in the hands of non-criminal institutions, are safer than their fellow patients in countries like the UK, which rely on the criminal law to make the prohibition stick.

It is hard for me to see how any reasonable person, surveying what we know about the ineffectiveness of control of medical behaviour by the criminal law, and the positive experiences to date with non-criminal control, could possibly come to any other conclusion that this: if we really want to improve patient safety, the control of medical end-of-life

[8] The Shipman Inquiry, First Report: *Death Disguised* (Chair Dame Janet Smith) (London: HMSO, 2002).

[9] R. Magnusson, *Angels of Death: Exploring the Euthanasia Underground* (New Haven, CT: Yale University Press, 2002).

practice should be taken out of the hands of the institutions of criminal law. Something better should be put in their place.

The argument for legalisation of euthanasia

My second thesis is that whether or not it is decriminalised, moral and legal consistency require that euthanasia be made legal. I shall assume that the other sorts of MBPSL remain legal, so all I need to show is that there are no relevant legal or moral differences between them.

We have seen that doctors intentionally cause the death of their patients much more frequently by administering lethal doses of pain relief or by withholding or withdrawing treatment that is essential to continued life, than they do by administering euthanatica. We have also seen that euthanasia, as practised in the Netherlands and Belgium, is a safer practice. Nevertheless, euthanasia is prohibited and the other two are not. What are the most important legal and ethical arguments for the different treatment?

One common argument used to distinguish withholding or withdrawing treatment from euthanasia is that abstention does not involve 'active' causing of death, but only a 'passive' failure to act.[10] But the idea that a 'mere omission' does not give rise to criminal liability does not apply in the case of abstention from treatment, for two reasons. In the first place, a 'mere' omission is hardly possible in the context of modern health care (one must turn off the ventilator, give a non-resuscitation order, instruct nursing staff not to intervene, and so forth). To convince oneself of the absurdity of the idea that abstention only involves an omission, one only has to think of the situation in which someone other than a doctor causes a patient to die by sneaking into his hospital room and turning off the ventilator. Does anyone seriously doubt that this is murder?

In the second place, freedom from criminal liability for an omission only applies if there is no duty to act, and a doctor definitely does have a duty to give his patient appropriate treatment. I shall not belabour the point further, since even John Keown, a persistent and vigorous opponent of legal euthanasia, has argued to the same effect; he rejects, as I do, a lot of rather loose language in the *Bland* case[11] – and many other places as well – as simply untenable.[12] It looks, in short, as if

[10] See the judgment in *Airedale Trust* v. *Bland* [1993] 1 All ER 821, for instance.
[11] Ibid.
[12] See J. Keown, *Euthanasia, Ethics and Public Policy* (Cambridge University Press, 2002), pp. 12–15.

almost a third of all deaths in the UK – deaths due to abstention – are *prima facie* at least as problematic as euthanasia. I shall return to this in a moment.

So much for 'mere' omissions. The common way of seeking to distinguish the legal treatment of euthanasia from the way that administration of lethal pain relief (and, in Keown's case, abstention from life-prolonging treatment as well) is treated, is the so-called doctrine of 'double effect'. A doctor who gives pain relief or withdraws or withholds treatment, knowing full well that doing this will hasten the patient's death, is thought not to be criminally liable so long as his *intent* was to relieve suffering or to put an end to burdensome treatment, and not to cause the patient to die.[13] That in the circumstances there was no way to do the one without the other is supposed to be irrelevant.

The doctrine of double effect, whatever its merits may be in a particular sort of moral theory, and however common such reasoning may be in daily life, has no place in the criminal law. The behaviour concerned in the cases that the doctrine seeks to distinguish – 'intentionally' as opposed to 'non-intentionally' causing death – is, as seen by an observer, indistinguishable (it consists in both cases of removing a ventilator, for example, or giving a lethal injection of pain relief). To make a distinction between acceptable and unacceptable causing of death by invoking the actor's subjective intention, we would have to know what that 'intention' is.

Let us consider the case of a person being hanged. The hangman puts a chair under his feet to support him while his head is put in the noose. The hangman then kicks the chair out from under his feet. Keown would, I think, agree that we cannot solve the problem of the hangman's criminal liability by trying to distinguish between the 'act' of hanging and the 'mere omission' to continue supporting the victim's feet with the chair. But what is the hangman's 'intention' – to cause the person to die or to remove the chair?

We might try to solve this problem the way Dutch and other researchers have sought to do. They distinguish between deaths due to euthanasia and to pain relief simply by asking the doctor concerned what his 'intention' was.[14] The example of the hangman demonstrates how artificial this is. Used in a criminal trial, such an approach would make liability dependent on confession. Nor would the confessions be reliable, since, as psychologists have demonstrated, actors do not reliably know what the intentions that lead to their actions are.[15]

[13] See, e.g., *R* v. *Adams* [1957] Crim LR 365.
[14] See Griffiths *et al.*, *Euthanasia and Law*, p. 151.
[15] See, e.g., D. Wegner, *The Illusion of Conscious Will* (Cambridge, MA: MIT Press, 2002).

Neither asking the person concerned about his subjective mental state, nor pretending to god-like omniscience is a reliable basis for deciding questions of criminal liability. This has been known for centuries to those involved in the administration of the criminal law. This is why, for purposes of criminal liability, 'intention' refers to something that can be known objectively and debated in court: a person is taken to intend the natural and probable results of his behaviour.

In the Netherlands, there was for some years considerable confusion about the legal status of the idea of double effect, due to the fact that those who discussed the legality of medical behaviour that causes death were practitioners of medical law and medical ethics. They were poorly informed about the criminal law. Deaths resulting from withholding treatment or administering pain relief were reported as 'natural' deaths and therefore did not reach the prosecutors and the courts. Recently, it has become clear that Dutch prosecutors never accepted the idea of double effect current in medical circles. The State Commission on Euthanasia (which included two very highly regarded prosecutors) explicitly rejected the idea in its report of 1985. In the most recent policy statement of the Royal Dutch Medical Association (concerning palliative sedation), the whole approach has been abandoned. What is critical for the legitimacy of potentially lethal treatment is not what the doctor thinks his subjective intention was, but whether what he did in the circumstances was *medically indicated*. Behaviour by a doctor that intentionally leads to the patient's death only falls within the justification available to doctors for what otherwise would be serious criminal offences if the behaviour was medically indicated in the circumstances. This was exactly the position of the State Commission.[16]

The foregoing considerations bring me to the following conclusion on the distinction between euthanasia and pain relief or abstention. All intentional causing of death – that is, all behaviour whose natural and probable consequence is the death of a person – is illegal, regardless of whether it can be described as an act or an omission, and regardless of the subjective intention of the actor – *unless*, that is, it can be *justified*.

The justification for withdrawing treatment is that considering the lack of any significant remaining benefit to the patient from the treatment itself, the invasion of bodily integrity involved is no longer justifiable. The treatment has become 'futile'. Futility, not omission or subjective intention, was the real ground for justifying the doctor's proposed

[16] For the history summarised in this paragraph and references to the reports of the State Commission and the Medical Association, see Griffiths *et al.*, *Euthanasia and Law*, pp. 64–6.

behaviour in the *Bland* case.[17] The justification for administering potentially lethal pain relief is that doing so is the only way to relieve severe suffering, which, in the circumstance of imminent death, is a more important consideration than postponing death. I conclude that in both the case of abstention and that of pain relief, it is the existence of a justification that accounts for the fact that the doctors who, in these two ways, are responsible for about half of all deaths every year, are generally not held criminally liable for doing so.[18]

Well, then: if it is the availability of a justification that is at issue, the case for legalising euthanasia seems quite simple. The fundamental principle on which the justifiability of pain relief and abstention rests weighs just as strongly in the case of euthanasia. And there is an additional powerful justification in the case of euthanasia (but not – at least, not necessarily – in the other two cases), namely, that the patient himself clearly, competently and consistently asks for it. I conclude that from the legal and moral perspective, abstention and pain relief are not significantly different from euthanasia. To the extent that there is a difference, the case for legalising euthanasia is stronger.

Apart from moral and legal consistency, there are some practical arguments in favour of legalisation of euthanasia. These stand the notorious *canard* of a 'slippery slope' on its head:[19] it is not only those who are allowed to die in a way that they themselves choose, but many others as well, whose safety at the hands of their doctors has been improved by the legalisation of euthanasia. The experience of regulating legalised euthanasia has had a spill-over effect on other forms of end-of-life practice. In countries where euthanasia or physician-assisted suicide is legal (Belgium, the Netherlands, Switzerland), doctors discuss all end-of-life decisions much more frequently with the patient and his or her family, and with other medical staff, than in non-permissive countries (the UK falls roughly halfway between the two groups).[20] Euthanasia takes place in the open, and is a subject of normal conversation between doctors and patients and their families.[21] In short, legalisation has been closely

17 See, e.g., Lord Goff in *Bland*, at 869–70.

18 For a much fuller discussion, see M. Otlowski, *Voluntary Euthanasia and the Common Law* (Oxford University Press, 1997).

19 See, further, Griffiths *et al.*, *Euthanasia and Law*, Chapter 19.

20 See ibid., p. 490.

21 See, e.g., F. Norwood, *The Maintenance of Life: Preventing Social Death through Euthanasia Talk and End-of-Life Care – Lessons from the Netherlands* (Durham, NC: Carolina Academic Press, 2009). Similarly, Paillet has shown that in France, where the use of lethal drugs to end the life of a newborn baby is completely illegal, the practice is quite common (see Griffiths *et al.*, *Euthanasia and Law*, p. 249), but this and other end-of-life decisions in neonatology are essentially never discussed with the parents or

associated with a generally more transparent medical practice and therefore with a lower level of social danger surrounding medical practice at the end of life.

The second development in the direction of increased patient safety – clearly influenced by the earlier experience with the legalisation (and decriminalisation) of euthanasia – is that the 'rules of careful practice' and the non-criminal control procedures created to make euthanasia practice safe have served as a model for the development of control over other sorts of MBPSL. In neonatology there is now a national protocol dealing with withholding and withdrawing treatment and active termination of life, a national review committee for cases of active termination, and a prosecution guideline that in effect defers judgment to the review committee.[22] There are currently moves in the direction of medical protocols for withdrawal of treatment in adult intensive care practice.[23] There is also a national protocol on the administration of palliative sedation.[24] The substantive provisions and the procedures provided for in these protocols borrow heavily from the control regime for euthanasia.

The limitation to 'an appropriate case'

If euthanasia is recognised as legal, what sort of limits should be imposed on the practice? I am not going to deal with procedural protections and *ex ante* or *ex post* review, because questions concerning these requirements have proven relatively easy to deal with.[25] Neither am I going to deal with the rather more difficult question of whether the practice should be limited to doctors; I shall simply assume that it is. I want to focus on what in the Netherlands has proven the hardest nut of all to crack: what conditions must the patient fulfil if it is to be justifiable that a doctor accede to his request for euthanasia?

Many opponents of legalisation make, in connection with what is often referred to as the 'conceptual' version of the 'slippery slope'

with other medical staff (A. Paillet, *Sauver la vie, donner la mort* (Paris: La Dispute/SNEDIT, 2007). By contrast, Moratti has shown that in the Netherlands, decisions to withdraw treatment are discussed by the whole medical team and are made together with the parents (S. Moratti, *Foregoing 'Medically Futile' Life-Prolonging Treatment in Dutch Neonatal Intensive Care Units* (New York: Edwin Mellen Press, in press).

22 See Griffiths *et al.*, *Euthanasia and Law*, Section 6.2.2.

23 See ibid., pp. 59–64. 24 See ibid., pp. 66–71.

25 The procedural protections of Dutch law are covered in Griffiths *et al.*, *Euthanasia and Law*, Chapter 4 *passim*. The gradual movement from *ex post* towards *ex ante* review is discussed in Section 4.2.4.4.

argument,[26] what I believe to be a simple mistake of ethical logic. They assume that a moral practice must be based on one fundamental ethical principle.[27] They then argue that if the justification of euthanasia is based on self-determination, the practice will sooner or later have to be extended to everyone who wants to die, and that current limitations to the sick and suffering will have to be abandoned. On the other hand, if the justification is based on beneficence, sooner or later doctors will have to be allowed to cause the death of anyone whom they think would be better off dead, whether or not the person has requested euthanasia.[28]

What this sort of doomsday thinking overlooks is that *both* principles are essential to a full justification of euthanasia. Self-determination justifies at most the legality of suicide that one carries out oneself. Euthanasia always requires an *additional* justification, namely for the involvement of *another* person. When that other person is a doctor,[29] the justification of his or her involvement in euthanasia will be limited to the sorts of cases in which the involvement of a doctor is appropriate at all – that is, cases in which the patient's request arises in the context of a doctor–patient relationship and the patient's wish to die is based on suffering that derives from a 'medical' condition.

It is precisely this limitation that has led to the two most difficult and controversial cases in the development of Dutch euthanasia law. Limiting access to legal euthanasia to 'patients' leaves some large groups of people out in the cold, people who have reasons we should respect for wanting to die, but who want this to happen in an effective, non-violent and humane way. In the *Chabot* case of 1994,[30] the woman involved had

[26] For explication and critique, see S. W. Smith, 'Fallacies of the Logical Slippery Slope in the Debate on Physician-Assisted Suicide and Euthanasia' (2005) 13 *Medical Law Review*, 224.

[27] The moral fallacy is similar to that exposed long ago by Hart in his essay on the moral justification of punishment. He argues that it is a mistake to think that this lies simply either in utilitarianism (general and special prevention) or in retribution. The 'general justifying principle' of punishment is utilitarian, but pursuit of this goal is limited by another principle, which requires individual blameworthiness. See H. L. A. Hart, 'Prolegomenon to the Principles of Punishment', in *Punishment and Responsibility: Essays in the Philosophy of Law* (Oxford University Press, 1968).

[28] Keown, *Euthanasia*, p. 123.

[29] The limitation to doctors is not a matter of principle but of effective control. It is less necessary in the case of assisted suicide, in particular when the practice is institutionalised and surrounded by appropriate controls, as in Switzerland. In such a situation, the presence of a 'medical' condition may no longer be necessary, but the involvement of the other person certainly does require justification that goes beyond self-determination.

[30] *Office of Public Prosecutions* v. *Chabot*, *Nederlandse Jurisprudentie* 1994 No. 656. See Griffiths *et al.*, *Euthanasia and Law in the Netherlands*, pp. 149–51 (discussion), 329–38 (court judgment); see also J. Griffiths, 'Assisted Suicide in the Netherlands: The *Chabot*

suffered some awful tragedies in her family life. In the opinion of the psychiatrist who ultimately assisted her suicide, she was not suffering from any psychiatric disorder, but the depth of her suffering and the single-mindedness of her wish to die were not questioned by anyone. The Dutch courts held that assistance with suicide can be justifiable in such a case. But some commentators, otherwise supportive of Dutch euthanasia policy, thought the courts had gone too far. The woman's problem was, in their view, not a 'medical' one, and the involvement of a doctor was therefore not justifiable.[31]

The more recent *Brongersma* case of 2003 concerned an elderly man who was treated by the courts as being simply 'tired of life'.[32] His doctor found that he was suffering unbearably and assisted with his suicide. The doctor was ultimately convicted, since the man's existential suffering was regarded as essentially non-medical in nature.

While the results in these two cases seem to point in rather different directions, they do make clear that the involvement of another person, specifically a doctor, sets limits on euthanasia practice, limits that cannot be accounted for simply in terms of self-determination. The result is that assistance with suicide is unavailable to many elderly people who desperately want an effective and humane way of ending a life that has become a meaningless burden to them.[33]

We have not seen the end of this issue in the Netherlands, any more than we have managed to deal with the trap confronting those in an early stage of Alzheimer's, who for good medical reasons dread what the future has in store for them, but who know that when that future arrives they will no longer be competent to ask for euthanasia. In short, even in the Netherlands we still have a long way to go in sorting out the problems associated with limiting justifiable euthanasia to an 'appropriate case'.

Conclusions

What can we conclude? *Decriminalisation* of euthanasia is important for reasons having to do with the safety of dying patients. *Legalisation* is

Case' (1995) 58 *Modern Law Review*, 232 and J. Griffiths, 'Assisted Suicide in the Netherlands: Postscript to *Chabot*' (1995) 58 *Modern Law Review*, 895.

31 See Griffiths *et al.*, *Euthanasia and Law*, pp. 122–3.

32 See Griffiths *et al.*, *Euthanasia and Law*, pp. 35–9, 123. In fact, the man had a number of physical ailments of the sort characteristic of advanced age, including incontinence and a disturbed sense of balance; his doctor had not considered it important to justify the decision to give assisted suicide in such 'medical' terms.

33 Hence the recent campaign by a citizens' action group demanding that individuals over seventy who are 'tired of life' should have the right to an assisted death (see www.uitvrijewil.nu).

called for because of the absence of any relevant differences between euthanasia and well-established legal ways in which doctors intentionally cause the deaths of patients but are considered to be justified in doing so. But when decriminalisation and legalisation have been accomplished, the really difficult ethical, legal and practical problems have only just begun.

3 Five flawed arguments for decriminalising euthanasia

John Keown

Introduction

There could hardly be a more apposite subject for a book on criminal law and bioethical conflict than euthanasia. The question of whether euthanasia should be decriminalised has been debated on both sides of the Atlantic for decades, and in recent years the debate has seriously intensified. This essay considers five of the main arguments standardly raised in favour of decriminalising voluntary, active euthanasia (VAE).

For the purposes of the essay, I shall adopt the definition of euthanasia employed by the House of Lords Select Committee on Medical Ethics, chaired by Lord Walton, which reported in 1994. It defined euthanasia as 'a deliberate intervention undertaken with the express intention of ending a life to relieve intractable suffering' (a definition that could reasonably be extended to include any reason that death is thought to benefit the patient). This essay's focus, therefore, is actions rather than omissions that shorten life. This is not to deny the possibility of intentional killing by omission, but simply to reflect the current euthanasia debate. Other Committee definitions adopted here concern the presence or absence of a request by the patient: euthanasia is 'voluntary' when carried out at the patient's request (VAE), 'non-voluntary' when the patient does not have the capacity to make a request (NVAE), and 'involuntary' when the patient is competent to make a request but does not do so (IVAE). The Committee defined physician-assisted suicide (PAS) as the provision by a doctor of help to commit suicide to a competent patient who has formed a desire to end his or her life.[1]

This and the preceding chapter were written after an oral debate between the two authors at the AHRC-funded conference, *Good, Bad or Indifferent? Medicine and the Criminal Process*, at the University of Manchester on 4 November 2009. However, neither has read nor reacted to the other's contribution to this volume.

[1] House of Lords Select Committee on Medical Ethics, *Report of the Select Committee on Medical Ethics* (HL Paper 21-I of 1993–94) (London: HMSO, 1994), paras. 20–6.

The Committee defended the criminal law's prohibition on intentional killing as the 'cornerstone of law and of social relationships', which 'protects each one of us impartially, embodying the belief that all are equal'.[2] The Committee was right to do so. The protection of innocent human beings, particularly from intentional attack, is a cardinal if not paramount purpose of the criminal law and is essential to the common good. The principle embodied by that prohibition, the 'sanctity' or 'inviolability' of human life, is a key component of any sound framework of human rights. As Lord Goff observed in *Bland*, the principle is 'fundamental' and has been 'long recognised not only in our own society but also in most, if not all, civilised societies throughout the modern world'.[3] The Committee was also right to conclude that if euthanasia were decriminalised it would not be possible to frame adequate safeguards against NVAE and that the vulnerable would feel pressure to request early death.[4] Let us now evaluate five of the main arguments for the decriminalisation of VAE/PAS.[5]

Hypocrisy of the criminal law

The first argument is that the criminal law is morally hypocritical. It prohibits doctors from intentionally administering a lethal injection, yet allows them to administer palliative drugs which they foresee may or will shorten life, and to withhold/withdraw life-prolonging treatment which they foresee may or will hasten death.

The argument is misconceived. Standard definitions of euthanasia rightly refer to an *intention* to shorten life. Intention is, in its ordinary English meaning of purpose, goal or aim, different from foresight. You may buy a lottery ticket intending to win, without foreseeing that you will. Conversely, you may foresee (whether as likely, virtually certain or certain) the discomfort of having a tooth extracted without intending it. The distinction is not only linguistic but moral. It is always wrong to bring about a bad consequence intentionally; it is not always wrong to do so foreseeably. If we make a bad consequence our aim, we endorse it and that endorsement corrupts our character. Though we can always avoid intending bad consequences, we cannot always avoid foreseeably bringing them about, by either our actions or our omissions. Bad side effects are often an inevitable part of our good purposes. Good doctors foresee the

[2] Ibid., para. 237. [3] *Airedale N.H.S. Trust* v. *Bland* [1993] A.C. 789, 863.

[4] House of Lords Select Committee on Medical Ethics, *Report*, paras. 238–239.

[5] For a fuller consideration of arguments against decriminalisation, see J. Keown, *Euthanasia, Ethics and Public Policy* (Cambridge University Press, 2002) and J. Keown, 'Against Decriminalising Euthanasia: For Improving Care' in E. Jackson and J. Keown, *Debating Euthanasia* (Oxford: Hart Publishing, 2012).

bad side effects of treatment; they never try to bring them about. The euthanasiast doctor who intentionally ends the patient's life judges that the patient's life is no longer worth living and turns from healer into killer. The doctor who aims to alleviate pain, merely foreseeing that life may be shortened as a side effect, does neither. Though we are morally *responsible* for the effects of our actions, good and bad, intended and foreseen, we may not be *blameworthy* for the foreseen bad effects of our conduct. The morality of our actions is not, then, simply about the effects we bring about: it is also about our state of mind in relation to those effects, not least our intention. This is why conviction for serious criminal offences standardly requires not simply that we have brought about a bad consequence, but that we have done so intentionally (by aiming to bring it about) or recklessly (by consciously taking an unjustified risk of bringing it about).

The principle of 'double effect' indicates when conduct is morally justified, even though it may or will bring about bad as well as good effects. The principle holds that it is permissible to bring about a foreseen bad consequence if the bad effect is not intended as an end or a means, and the foreseeable causing of the side effect does not violate other moral norms, especially fairness. Double effect has long been adopted by professional medical ethics and the criminal law. In the *Bland* case, Lord Goff referred to:

> the established rule that a doctor may, when caring for a patient who is, for example, dying of cancer, lawfully administer painkilling drugs despite the fact that he knows that an incidental effect of that application will be to abbreviate the patient's life.[6]

On an earlier occasion his Lordship had made it clear that murder requires an intent to kill, and that while foresight can be evidence of intention, it is not the same as intention.[7] When, in 1997, the then Attorney-General, Lord Williams of Mostyn, was asked whether the government was satisfied with the current state of the criminal law and medical practice in relation to the palliative treatment of terminally ill patients, he replied that the government *was* satisfied. Citing Lord Goff's statement in the *Bland* case, Lord Williams commented that he did not think doctors had the slightest difficulty understanding the distinction between intending death and merely foreseeing death. He added that the law was neither difficult nor obscure; it was 'perfectly plain'.[8] Although it

[6] *Bland*, 867.

[7] R. Goff, 'The Mental Element in the Crime of Murder' (1988) 104 *Law Quarterly Review* 30.

[8] Hansard, HL, vol. 583, cols. 742–744, 20 November 1997.

is true to say that the English courts have sometimes come close to equating intention with foresight of virtual certainty, they have, as Professor Skegg has pointed out, 'tended to say that foresight of virtual certainty is something from which intention may be found or inferred, and . . . have stopped short of saying that such foresight is itself a form of intent'.[9] The Dutch, the pioneers of VAE/PAS, hold that euthanasia requires an intention to shorten life. Their national surveys of medical practice at the end of life, which we shall mention, have consistently defined euthanasia as involving an intent or purpose to shorten life, not merely the foreseen shortening of life.

In short, Anglo-American law and professional medical ethics draw an important distinction, recognised by sound philosophy and common sense, between the intended and the merely foreseen hastening of death. To conflate euthanasia with palliative treatment or with withholding/withdrawing life-prolonging treatment is to elide a crucial moral and legal distinction. If there were no distinction, there would be no euthanasia debate. The law and medical practice are not hypocritical: they are Hippocratic.

Autonomy, VAE and NVAE

A second argument is that decriminalising VAE is required by respect for autonomy and that, as only autonomous patients could request it, there is no risk of a 'slippery slope' to NVAE. Decriminalising VAE would be a discrete step involving no endorsement of NVAE.

This argument prompts several questions. What, precisely, counts as an 'autonomous' request? How 'autonomous' would requests be in practice? Is respect for autonomous choice a moral principle, and, if so, why? If it is, is it not trumped by the principle of the inviolability of life that prohibits intentional killing, a prohibition described (we shall recall) as the 'cornerstone of law and of social relationships'?[10] Here, however, we shall focus solely on the question of whether there is a logical slippery slope from VAE to NVAE. (This question tends to be overshadowed by the question, albeit scarcely less important, of whether there is an empirical slippery slope, which largely concerns the practical difficulty of policing the line between VAE and NVAE.)

Despite the emphasis euthanasia campaigners place on autonomy, the case for VAE rests at least as much on the belief that some patients

[9] P. D. G. Skegg, 'Medical Acts Hastening Death' in P. D. G. Skegg and R. Paterson (gen. eds.), with J. Manning, J. Dawson, N. Peart and W. Brookbank, *Medical Law in New Zealand* (Wellington: Thomson Brookers, 2006), pp. 505, 524. See, e.g., *R* v. *Woollin* [1999] 1 A.C. 82. (HL).

[10] House of Lords Select Committee on Medical Ethics, *Report*, para. 237.

would be (because of terminal illness, suffering or whatever) 'better off dead'. Campaigners are not proposing that VAE/PAS should be available simply because the patient has made an autonomous request. According to their own standard proposals, doctors would have to decide whether the patient's autonomous request should be granted. And how would the doctor decide *other than on the basis of a belief that death would benefit the patient?* And a doctor who is prepared to make such a judgement forfeits any principled objection to NVAE. For if death would benefit the patient, why deprive the patient of that benefit merely because the patient cannot request it? It is sometimes argued in reply that in the case of VAE, it is the patient's subjective judgement that life is no longer worth living that justifies the doctor in ending the patient's life. But no responsible doctor is going to kill a patient merely because the patient judges, however autonomously, that death would be a benefit. A responsible doctor will make his or her own independent judgement whether that criterion is satisfied (even though in doing so the doctor may well take the patient's opinion into account). Illustrating the decisive role of the doctor, Dutch doctors reject many autonomous requests for euthanasia. In short, the patient proposes, but the doctor disposes.

Professor John Griffiths *et al.* have challenged this logical slope argument. They rightly note that in VAE a second person, the doctor, is involved and that the doctor requires more to justify VAE than the patient's request. The justification, they add, lies in the doctor's duty to relieve suffering. But why may this same duty not justify NVAE? Griffiths *et al.* answer: 'The requirement of a request will not be swept away precisely because it is based on an entirely different principle.'[11] The principle of respect for autonomy is indeed a different principle, but it is irrelevant in the case of an incompetent patient (at least if the patient has not left an advance directive). As we shall see, just as Dutch courts invoked the doctor's duty to alleviate suffering to justify VAE, they have more recently invoked the very same duty to justify NVAE. Discussing the Dutch courts' endorsement of infanticide, Griffiths *et al.* write:

> The applicable norms in the Netherlands have assuredly changed in the direction of open acceptance of the legitimacy of termination of life of severely defective newborn babies ... [T]he influence on these changes of the way euthanasia had earlier been legalised and regulated is obvious. In this sense, one might speak of a normative slippery slope.[12]

[11] J. Griffiths, H. Weyers and M. Adams, *Euthanasia and Law in Europe* (Oxford: Hart, 2008), p. 514; see also p. 49.

[12] Ibid., p. 252.

The justification invoked by the Dutch courts for their endorsement of NVAE precisely tracks the logical argument that even if autonomy is not in play, beneficence is. The courts' reasoning, far from being a refutation of the logical argument, is a vivid illustration of it. The rejection of the logical slope by Griffiths *et al.* is therefore puzzling. It is, indeed, doubly puzzling, for they appear, like many supporters of VAE, to have no objection to NVAE. They even suggest that both VAE and NVAE should be classified as 'normal medical practice', that doctors should be allowed to report them confidentially to local committees composed of fellow doctors, and that the reports should be kept secret from prosecutors, even if they disclose murder.[13]

As the logical slope argument shows, permitting VAE is no 'private' matter concerning only a few highly autonomous patients. It clearly has profound ramifications for the wellbeing of *all* patients who might be judged 'better off dead', not least those who are elderly, dying, have dementia or disabilities. Small wonder that disability groups on both sides of the Atlantic are at the forefront of opposition to VAE/PAS.

The failure of the criminal law

A third argument is that the current criminal prohibition of VAE/PAS is a failure, that they are commonly practised, and that decriminalisation would 'bring them out into the open' and permit effective legal regulation.

First, there appears to be little evidence that the law against VAE/PAS is frequently broken, let alone that the law is a dead letter. In 2006, Professor Clive Seale published the results of his survey of end-of-life decisions by UK doctors. Seale found that the proportion of deaths from PAS, VAE and NVAE was 'extremely low'.[14] By contrast, as Griffiths *et al.* admit, the Dutch rate of euthanasia and physician-assisted suicide is by far the highest in Europe. The comparable rates in 2001–2 between the Netherlands and the UK, expressed as a percentage of all deaths, were:

[13] J. Griffiths, A. Bood and H. Weyers, *Euthanasia and Law in the Netherlands* (Amsterdam University Press, 1998), pp. 286–93. See also Griffiths *et al.*, *Euthanasia and Law*, p. 73, n. 88.

[14] C. Seale, 'National Survey of End-of-Life Decisions Made by UK Medical Practitioners' (2006) 20 *Palliative Medicine* 3, 6.

Table 3.1. *Rates of euthanasia and physician-assisted suicide in the Netherlands and the UK, 2001–2*

	As a % of all deaths	
Type of euthanasia/PAS	Netherlands	UK
VAE	2.59	0.16
PAS	0.21	0.00
NVAE	0.60	0.33

Adapted from Griffiths, Weyers and Adams, *Euthanasia and Law*, p. 489, Table 17.2

A repeat survey in the UK by Seale, published in 2009, concluded: 'Euthanasia, physician-assisted suicide and the ending of life without an explicit patient request ... are rare or non-existent at both time points.'[15] The second survey also found, as a result of more refined questions, that 'non-treatment decisions' and 'double effect measures' were 'much less common than suggested in earlier estimates, rarely involving intent to end life or being judged to have shortened life by more than a day'.[16] Seale did find a high rate of 'continuous deep sedation' in the UK, but his survey did not ask doctors whether they used it with any intent to hasten death, and he recommends further research into the context in which it is practised. Griffiths *et al.* suggest that in the Netherlands there is evidence that doctors are using 'terminal sedation' (defined as continuous deep sedation combined with withholding/withdrawing artificially delivered food and fluids) to end life so as to avoid the law's requirements for VAE.[17] Even so, it was reported that deaths by VAE rose 13 per cent in 2009 to 2,636, following an increase of 10 per cent in 2008, and that the 2008 increase had prompted the health ministry to set up an investigation.[18]

Further, Griffiths *et al.* suggest a sizeable incidence of illicit assistance in suicide in the Netherlands by doctors and laypeople. They cite research by Dr Chabot suggesting that some 1,600 people per year have committed assisted suicide outside the legal guidelines, by taking sleeping pills that were supplied by a doctor or prepared for ingestion by a layperson, or both. They observe that many serious crimes

[15] C. Seale, 'End-of-Life Decisions in the UK Involving Medical Practitioners' (2009) 23 *Palliative Medicine* 198, 201.

[16] Ibid., 198. [17] Griffiths *et al.*, *Euthanasia and Law*, p. 71.

[18] 'Netherlands: Sharp Growth in Euthanasia Deaths' (see www.euthanewsia.ca/archive/2010/06/enid201006173840.4.html).

are therefore involved over which control is essentially non-existent.[19] That the incidence of VAE/PAS should increase with decriminalisation accords with common sense. Dr Neil Gorsuch, author of a major treatise on VAE/PAS and now a judge on the US Federal Court of Appeals, has pointed out that such an increase also accords with the 'law of demand', which holds that, other things being equal, the quantity demanded of a good falls as its cost rises. He explains:

> Consistent with the law of demand, one would expect that if certain 'costs' associated with assisted suicide and euthanasia (for example, the social stigma and difficulty of finding a willing physician to help when the practices remain illegal) are lowered or eliminated by legalization, *more*, not fewer, people would take advantage of this fact and seek an early death.[20]

In relation to the decriminalisation of abortion, the statistics in jurisdictions like the UK and the USA are certainly consistent with the law of demand. Professor Glanville Williams, an ardent pro-choice campaigner, wrote that decriminalisation produced an 'entirely new clientele', and that its effect was 'to add the total of legal abortions to the total of illegal abortions rather than to reduce the number of illegal abortions'.[21] Why should it be any different with the decriminalisation of VAE?

Second, even if VAE/PAS were frequently performed in countries like the UK where it is illegal, would this amount to a cogent argument for decriminalisation? For all we know, female genital mutilation is frequently practised by disreputable doctors in the UK. Would this justify the repeal of the Female Genital Mutilation Act 2003? Perjury may also be frequent. Speeding is very frequent. Why is frequency not a reason for more effective enforcement rather than for decriminalisation? Moreover, whatever the incidence of illegal VAE/PAS in countries like the UK and the USA, the laws against them are clearly a long way from laws like Prohibition, which have not only failed to eliminate the proscribed conduct, but have been so widely and publicly flouted by respectable members of society as to lose 'even the character of symbolising a real societal commitment to the values they purport to uphold'.[22]

Third, would decriminalisation of VAE/PAS bring whatever illegal practice there may be 'out into the open'? Why should we assume that any doctors who are currently willing to commit the crime of murder

[19] Griffiths *et al.*, *Euthanasia and Law*, p. 184.
[20] N. M. Gorsuch, *The Future of Assisted Suicide and Euthanasia* (Princeton University Press, 2006), p. 133 (original emphasis).
[21] See J. Keown and D. Jones, 'Surveying the Foundations of Medical Law' (2008) 16(1) *Medical Law Review* 85, text at nn. 187–8.
[22] J. Finnis, 'Abortion and Legal Rationality' (1970) 3 *Adelaide Law Review* 431, 436.

and/or assisting suicide would be willing to comply with guidelines for VAE/PAS? It is claimed, however, that the experience in the Netherlands, where VAE/PAS were decriminalised in 1984, shows that legal 'safeguards' can ensure effective control and prevent any slide down the slippery slope. Let us now evaluate this fourth argument for decriminalisation.

Lessons from Dutch criminal law and medical practice

In 1984, the Dutch Supreme Court held that a doctor may claim the defence of 'necessity' in cases where, according to 'responsible medical opinion', the duty to alleviate suffering overrides the duty not to kill. Shortly after the decision, the Royal Dutch Medical Association issued guidelines for doctors.[23] In 2002, the Termination of Life on Request and Assisted Suicide (Review Procedures) Act came into force. In essence, the Act enshrines the guidelines in statutory form. Inter alia, the Act requires, first, that the patient made a 'voluntary and carefully considered' request (often described as an 'explicit request') for VAE/PAS; second, that the patient was experiencing 'unbearable suffering with no prospect of improvement'; and third, that before carrying out VAE/PAS, the doctor consulted with an independent doctor, and afterwards reported the case to the local medical examiner. The report is then to be forwarded by the examiner to a Regional Review Committee.

Defenders of the Dutch experience (not least the Dutch) claim that it shows that VAE/PAS can be effectively controlled. This claim is not obviously borne out by the evidence, much of which has been helpfully generated by the Dutch themselves in the form of government-sponsored national surveys into end-of-life medical practice in 1990, 1995, 2001 and 2005. In what follows, we shall rely largely on the evidence as it is set out in the work of John Griffiths *et al.*, leading defenders of Dutch euthanasia.[24]

Voluntary request In 2006, the Royal College of Psychiatrists noted a strong association in the terminally ill between a desire for a hastened death and clinical depression, and observed that once their depression was treated, 98–99 per cent changed their mind about wanting to die.[25] In the Netherlands, only 3 per cent of euthanasia cases involve

[23] Keown, *Euthanasia*, pp. 83–6.

[24] Griffiths *et al.*, *Euthanasia and Law*; Griffiths *et al.*, *Euthanasia and Law in the Netherlands*.

[25] *Statement from the Royal College of Psychiatrists on Physician-Assisted Suicide* (2006), para. 2.4 (see www.rcpsych.ac.uk/pdf/14.07.11%20Enc%2005.pdf). At time of writing, the Statement is under review.

psychiatric evaluation;[26] there must, therefore, be a question about the possible role of mental illness in the requests made by the remaining 97 per cent. Moreover, a Dutch survey of cancer patients found, to the surprise of the researchers, not only that depressed patients are more than four times as likely to request euthanasia (no fewer than 44 per cent of the depressed patients surveyed requested euthanasia), but that there was no significant association between depression and prognosis: it was the depressed patients, not the sickest patients, who seemed to be requesting euthanasia.[27]

Apart from the many cases raising doubts about how autonomous the request for VAE/PAS was, there have been many cases in which Dutch doctors have admitted that there was no explicit request at all. The Dutch data have disclosed that such cases of what is in Dutch law murder are not uncommon and are rarely prosecuted. The first Dutch survey revealed that in 1990 there were 2,300 cases of VAE and 400 cases of PAS. (These figures are conservative: they omit many other cases in which doctors admitted conduct, active and well as passive, in which their primary intention was to end life.) Remarkably, the survey also revealed that a further 1,000 patients (most of whom were incompetent, but 140 of whom were wholly competent and a further 110 partly competent) had been given a lethal injection without having made an explicit request. (By 2005, the number had halved, but each case remains one of murder.[28]) Griffiths *et al.* note that Dutch law may even be moving to permit such cases of NVAE. In the *Van Oijen* case, the defendant doctor was convicted of murdering a woman who was not suffering unbearably and who had made it clear she did not want to die. (Reflecting the extreme leniency of the Dutch courts, his sentence was one week's imprisonment, suspended for two years.) Griffiths *et al.* observe: 'the Supreme Court ruled that the justification of necessity in principle can be available in a case of ending the life of a dying patient without the patient's request'.[29] They write:

> If one may venture a prediction, it would be this: the idea of an 'inhumane death' will, in one form or another, come increasingly to be accepted not only as a legitimate (prospective) reason for requesting euthanasia but as a justification for 'help in dying' [NVAE].[30]

[26] Griffiths *et al.*, *Euthanasia and Law*, p. 174.

[27] M.L. van der Lee, J.G. van der Bom, N.B. Swarte, A.P.M. Heintz, A. de Graeff and J. van den Bout, 'Euthanasia and Depression: A Prospective Cohort Study Among Terminally Ill Cancer Patients' (2005) 23 *Journal of Clinical Oncology* 6607.

[28] See, generally, Keown, *Euthanasia*, Part III, and J. Keown, *Considering Physician-Assisted Suicide* (London: Carenotkilling Alliance, 2006), Part I, p. 1.

[29] Griffiths *et al.*, *Euthanasia and Law*, p. 41, citing *Nederlandse Jurisprudentie* 2005, No. 217: 9.

[30] Griffiths *et al.*, *Euthanasia and Law*, p. 73; see also p. 142.

They also note that Dutch Courts of Appeal have already endorsed NVAE in the case of disabled babies. Paediatricians have drawn up guidelines, notably the 'Groningen Protocol', to regularise this practice of infanticide.[31] Griffiths *et al.* reasonably suggest, in the light of these developments, that Dutch law is slowly but steadily moving in the direction of explicit recognition of a doctor's duty to ensure that his patient dies a 'humane' or 'dignified' death.[32] The ease with which such an arbitrary notion could be extended to vulnerable groups, including those who are elderly, have dementia or disabilities, scarcely needs emphasis.

Yet further evidence of the growing acceptance of NVAE is the striking statement by the lead authors of the Dutch national surveys that it is now the responsibility of *patients* to make it clear, when competent, if they do *not* want to be given a lethal injection should they become incompetent. As Dr Fenigsen has translated it:

> [I]t is the patient who is now responsible in the Netherlands for avoiding termination of his life; if he does not wish to be killed by his doctor then he must state it clearly orally and in writing, well in advance.[33]

'Unbearable suffering with no prospect of improvement' The requirement of 'unbearable suffering' has proved somewhat elastic. First, provided that a patient's assertion that suffering is unbearable is 'understandable' to a normal doctor, and provided it is due to a recognised medical disorder, it is largely a matter of a patient's subjective feeling. Second, the suffering need not arise from a terminal or even a somatic illness. This was so held by the Supreme Court in the *Chabot* case, where psychiatrist Dr Chabot helped a woman to kill herself solely because of her grief at the death of her two sons.[34] Third, fear of future dementia can qualify, even though the patient will not suffer at all when demented. If fear of dementia, why not fear of physical deterioration or dependency? Fourth, although the Supreme Court in the *Brongersma* case[35] ruled out existential suffering or 'tiredness of life', on the ground that the suffering must be predominantly due to a medically classified disease or disorder if a doctor is to be acting within the scope of professional competence, it is doubtful whether this limitation will stand. An

[31] See generally ibid., Chapter 6. For another sympathetic account of this development, see S. Moratti, 'End-of-Life Decisions in Dutch Neonatology' (2010) 18(4) *Medical Law Review* 471.
[32] Griffiths *et al.*, *Euthanasia and Law*, p. 143.
[33] Quoted in Keown, *Considering Physician-Assisted Suicide*, p. 6.
[34] *Office of Public Prosecutions* v. *Chabot*, *Nederlandse Jurisprudentie* 1994 No. 656, Supreme Court.
[35] *Nederlandse Jurisprudentie* 2003, No. 167.

advisory committee of the Royal Dutch Medical Association concluded in 2004 that euthanasia for existential suffering should be permitted.[36] In 2011, the Association concluded that 'vulnerability – extending to such dimensions as loss of function, loneliness and loss of autonomy – should be part of the equation' that physicians use to assess requests for euthanasia.[37] Provided that a patient's request is 'motivated in part by a medical ground',[38] the patient is eligible for euthanasia. But what is to prevent the patient (with or without the physician's collusion) misrepresenting an existential motivation as partly medical?

Reporting The four national surveys have shown that in thousands of cases doctors have flouted the legal requirement to report. In 1990, the rate of non-reporting was 80 per cent. By 2005, it had fallen but was still running at 20 per cent.[39] Each failure to report not only breaches the guidelines, but also involves the criminal offence of falsifying a death certificate by entering death by 'natural causes'.

Critics It is not surprising, in view of the above evidence, that the Dutch experience has drawn criticisms from many quarters, including the UN Committee on Human Rights. In 2001, the Committee ventilated several concerns, not least the NVAE of disabled babies. In 2009, the Committee reiterated concerns, not only 'at the extent of euthanasia and assisted suicides', but also that a physician can terminate a patient's life 'without any independent review by a judge or magistrate to guarantee that this decision was not the subject of undue influence or misapprehension'.[40] It recommended that the Dutch legislation be reviewed in light of the UN's recognition of the right to life. Critics include those who support VAE in principle. In 1998, Griffiths *et al.*, having noted the 'intrinsic ineffectiveness of control based on self-reporting',[41] observed that the control regime did not provide 'effective control', that the results of the first two national surveys that

[36] KNMG, *Op zoek naar normen voor het handelen van artsen bij vragen om hulp bij levensbeëindiging in geval van lijden aan het leven: verslag van de werkzaamheden van een commissie onder voorzitterschap van prof. J Dijkhaus* (Utrecht: KNMG, 2004).

[37] KNMG, *The Role of the Physician in the Voluntary Termination of Life* (Utrecht: KNMG, 2011), p. 22.

[38] Ibid., p. 25. If refused euthanasia, patients have a right to medical help to thirst to death; ibid., pp. 34–6.

[39] Griffiths *et al.*, *Euthanasia and Law*, p. 199.

[40] UN Human Rights Committee, 'Consideration of Reports Submitted by States Parties under Article 40 of the Covenant', CCPR/C/NLD/CO/4 (25 August 2009) para. 7 (see www.unhcr.org/refworld/publisher, HRC, 4aa7aa642,0.html).

[41] Griffiths *et al.*, *Euthanasia and Law in the Netherlands*, p. 257.

had by then been published had been 'pretty devastating', and that the regime seemed 'all bark and no bite'.[42]

In 2008, by contrast, they claimed that there is now a 'well-developed system of control, consisting of expert consultation before euthanasia is carried out and assessment afterwards by the Regional Review Committees, which have been designed to be acceptable to doctors'.[43] This about-face, which appears to rest on improvements in consultation and reporting, is puzzling. First: consultation. They point out that in 1999 a national programme (SCEN: Support and Consultation Euthanasia Netherlands) was instituted, under which doctors could be trained in the medical, ethical and legal aspects of euthanasia consultation, and that now in 90 per cent of reported cases doctors have consulted an SCEN physician. However, while this belated training programme may well have improved the quality of the consultation process in many cases, the requirement of consultation with an independent doctor is nothing new. Moreover, even consulting a doctor who has undergone an SCEN training course (and consultation with such a doctor is not legally required) is a long way from consulting a specialist in psychiatry or palliative care. As we noted above, only 3 per cent of euthanasia cases involve referral to a psychiatrist. Moreover, Dr van Oijen, who was convicted of murdering a patient against her express wish, was himself an SCEN consultant. Further, a consultant's failure to see the patient, a negative opinion by a consultant, or a failure to consult an independent consultant, will not necessarily trigger prosecution.[44] Although the legislation requires consultation, Griffiths *et al.* point out that the Review Committees interpret this in a 'flexible way', thereby correcting what they (oddly) describe as a 'legislative mistake'.[45]

Second: reporting. Griffiths *et al.* point to the introduction of Review Committees in 1998 – which replaced prosecutors as the destination for physicians' reports – and the subsequent improvement in the reporting rate from 20 per cent to 80 per cent. They contend that one of the most important advantages of these committees is that their decision-making is more transparent than the previous prosecutorial review system.[46] But greater openness about consideration of the reports has little bearing on the reliability of those reports. The system of review still relies on self-reporting, whose 'intrinsic ineffectiveness' they rightly criticised in 1998. Not surprisingly, the committees (staffed by a lawyer, doctor and ethicist, who mostly 'do their committee work on the side, being primarily active professionals'[47]) 'very rarely' find that the doctor has

[42] Ibid., p. 268. [43] Griffiths *et al.*, *Euthanasia and Law*, p. 49.
[44] Ibid., pp. 93–9, nn. 183–211. [45] Ibid., p. 84, n. 139.
[46] Ibid., p. 137. [47] Ibid., p. 138.

breached the requirements and refer the case to the prosecutors.[48] The prosecutors themselves 'hardly ever bring criminal charges in the tiny fraction of all cases that do come to their attention'.[49] Most reports received by the committees are 'unproblematic' and 'hardly afford anything to discuss'.[50] Moreover, the committees have found that doctors acted properly even when their actions 'were not entirely in conformity with the legal requirements'.[51] Of the few cases that they have referred to prosecutors, none has resulted in prosecution. The 'handful' of prosecutions in recent years are of non-reported cases, which 'happen accidentally' to come to prosecutorial attention.[52] And there are many unreported cases. For example, Griffiths *et al.* admit that although there are known to be 'a considerable number' of cases of NVAE, 'hardly any' have been reported.[53] In relation to the NVAE of infants, they write that 'control over termination of life of severely defective newborn babies cannot yet be regarded as legally adequate', and 'because of an apparently low reporting rate, it is certainly not effective'.[54]

How can a system of review by part-timers who rely on what they are told by those who report, and who are not told anything by those who do not, sensibly be described as a 'well-developed system of control'? As Judge Gorsuch has asked of the Dutch regime:

> Does a regime dependent on self-reporting by physicians who have no interest in recording any case falling outside the guidelines adequately protect against lives taken erroneously, mistakenly, or as a result of abuse or coercion? How would we ever know?[55]

Griffiths *et al.* suggest that to encourage doctors to report NVAE, it should be classified as 'normal medical practice'. Gorsuch notes that, curiously, Griffiths does not give any significant consideration to the question of whether allowing doctors to kill patients without consent might also lead to additional cases of abusive, coercive and mistaken killings, and that Griffiths' proposal would appear to rule out prosecution whether the doctor's motives were compassionate or cruel. Gorsuch adds:

> In Griffiths' preferred regime, only professional and civil sanctions would be available as remedies when doctors kill without consent – and even these remedies would be available only if and when doctors kill in the absence of what he calls 'normal medical practice' – although Griffiths fails to specify when he thinks killing a patient without consent should be considered 'normal'.[56]

[48] Ibid., p. 145. [49] Ibid. [50] Ibid., p. 136, n. 388. [51] Ibid., p. 135.
[52] Ibid., p. 128. [53] Ibid., p. 127. [54] Ibid., p. 252.
[55] Gorsuch, *The Future*, p. 114. [56] Ibid., p. 115.

Judge Gorsuch also points out that there is no guarantee that decriminalisation of NVAE would increase reporting: doctors might still be deterred by possible civil and professional sanctions. In short, Griffiths *et al.* appear to exaggerate the significance of changes that fail to ensure that the guidelines are satisfied in any given case, to meet the serious concerns that have been voiced by critics like the UN Human Rights Committee, or to remedy the 'intrinsic ineffectiveness' of a system based on self-reporting.

In 2004, Raphael Cohen-Almagor, a supporter in principle of VAE/PAS, who has conducted empirical research in the Netherlands, concluded that the Dutch regime suffers from serious flaws, that all the guidelines are regularly broken, and that the Dutch tend to condone NVAE. The high number of unreported cases, and the fact that many physicians did not wish to be bothered with the procedures, was, he found, 'alarming'.[57]

Moreover, in 2009, Dr Els Borst, perhaps the leading architect of Dutch euthanasia, reportedly told Dr Anne-Marie The, an expert on euthanasia, that legalised euthanasia had been followed by a decline in the quality of care for the terminally ill and that more should have been done to protect those who wanted to die naturally. Dr Borst said that euthanasia had been legalised 'far too early', adding: 'In the Netherlands, we first listened to the political and societal demand in favour of euthanasia. Obviously this was not in the proper order.' Dr The reportedly commented that palliative care was so inadequate in the Netherlands that patients often ask for euthanasia 'out of fear' of dying in agony.[58] So much, then, for claims that the Dutch have subjected VAE/PAS to effective control. The evidence generated by the Dutch themselves discloses widespread breaches of the guidelines, and with virtual impunity.

VAE for the 'non-vulnerable'

It is sometimes argued, fifth, that decriminalisation need involve no threat to the vulnerable because it could be confined to those who are clearly not vulnerable. In the *Pretty* case, Dianne Pretty asked the courts to order the Director of Public Prosecutions (DPP) to guarantee that he would not prosecute her husband if he assisted her to commit suicide. She argued that whatever the need might be for legal protection of the vulnerable, she was not vulnerable and that there was therefore

[57] R. Cohen-Almagor, *Euthanasia in the Netherlands* (Dordrecht: Kluwer Academic Publishers, 2004), pp. 175–9.

[58] S. Caldwell, 'Now the Dutch Turn Against Legalised Mercy Killing', *Daily Mail*, 9 December 2009.

no justification for the law's blanket prohibition. The Senior Law Lord, Lord Bingham, responded:

Beguiling as that submission is, Dr Johnson gave two answers of enduring validity to it. First, 'Laws are not made for particular cases but for men in general.' Second, 'To permit a law to be modified at discretion is to leave the community without law. It is to withdraw the direction of that public wisdom by which the deficiencies of private understanding are to be supplied.[59]

To Pretty's argument that the law unjustly discriminated against those who, like her, were physically unable to commit suicide without assistance, Lord Bingham responded that the criminal law could not be criticised as discriminatory because it applied to everyone:

Although in some instances criminal statutes recognise exceptions based on youth, the broad policy of the criminal law is to apply offence-creating provisions to all and to give weight to personal circumstances either at the stage of considering whether or not to prosecute or, in the event of conviction, when penalty is to be considered. The criminal law does not ordinarily distinguish between willing victims and others ... Provisions criminalising drunkenness or misuse of drugs or theft do not exempt those addicted to alcohol or drugs, or the poor and hungry.

He added:

'Mercy killing', as it is often called, is in law killing. If the criminal law sought to proscribe the conduct of those who assisted the suicide of the vulnerable, but exonerated those who assisted the suicide of the non-vulnerable, it could not be administered fairly and in a way which would command respect.[60]

Citing sources including the Walton Committee, which we recall unanimously rejected the case for relaxation of the law for reasons of both principle and prudence, his Lordship concluded that there were 'ample grounds to justify the existing law and the application of it'.[61] Lord Steyn observed that the law's total prohibition on assisting suicide was 'a legitimate, rational and proportionate response to the wider problem of vulnerable people who would otherwise feel compelled to commit suicide'.[62] On appeal, the European Court of Human Rights agreed with the Law Lords that a blanket ban did not breach the European Convention on Human Rights:

Doubtless the conditions of terminally ill individuals will vary. But many will be vulnerable and it is the vulnerability of the class which provides the rationale for

[59] *R (on the application of Pretty)* v. *DPP* [2002] 1 A.C. 800, [29]. [60] Ibid., [36].
[61] Ibid., [30]. [62] Ibid., [63].

the law in question. It is primarily for States to assess the risk and the likely incidence of abuse if the general prohibition on assisted suicides were relaxed or if exceptions were to be created. Clear risks of abuse do exist, notwithstanding arguments as to the possibility of safeguards.[63]

More recently, the European Court has held that states are under no obligation to facilitate assisted suicide. Upholding a Swiss law that requires a prescription in order to obtain lethal drugs, the Court noted that the risks of abuse if a state permitted assisted suicide could not be underestimated.[64]

Regrettably, the European Court indicated in *Pretty* that the UK's blanket ban did engage Dianne Pretty's right to respect for 'private and family life' guaranteed by Article 8(1) of the Convention, though the ban was saved by Article 8(2), which allows interference with that right if the interference is in accordance with law and necessary in a democratic society in the protection of certain interests, including the rights and freedoms of other people. The Court's interpretation of 8(1) was unsound. How can assisting suicide, a matter of the gravest *public* concern, ever be a *private* matter? The Court should have followed Lord Bingham, who pointed out that Article 8(1) sought to protect certain choices while people were living their lives, not the choice to live no longer.[65]

No less regrettably, the Law Lords have now gone even further than the European Court in misinterpreting Article 8(1). In the *Purdy* case, they went so far as to order the DPP to issue guidance spelling out the factors he would take into account in deciding whether to prosecute Debbie Purdy's husband should he assist her to commit suicide. Their decision was, with respect, unsound if not unconstitutional. As the Court of Appeal had ruled, making such an order 'would, in effect, recognise exceptional defences to this offence which Parliament has not chosen to enact'.[66] Indeed, Parliament has more than once chosen not to enact such exceptions. The Law Lords, by making the order, undermined Parliament's blanket ban on assisting suicide. So too have

[63] *Pretty* v. *United Kingdom* (2002) 35 EHRR 1, [74].

[64] 'En particulier, la Cour considère que l'on ne saurait sous-estimer les risques d'abus inhérents à un système facilitant l'accès au suicide assisté.' *Affaire Haas* c. *Suisse* (2011) [58] (see http://cmiskp.echr.coe.int/tkp197/view.asp?action=html&documentId=880260&portal=hbkm&source=externalbydocnumber&table=F69A27FD8FB86142BF01C1166DEA398649).

[65] *R (on the application of Pretty)* v. *DPP* [2002] 1 A.C. 800, [61].

[66] *R (on the application of Purdy)* v. *Director of Public Prosecutions* [2009] EWCA Civ 92, [79]; [2009] UKHL 45. See J. Keown, 'In Need of Assistance?' (2009) 159 *New Law Journal* 1340.

the guidelines subsequently issued by the DPP, which send the clear signal that if they are followed, there will be no prosecution.[67]

Conclusions

We have considered five major arguments in favour of decriminalising VAE/PAS. Despite the superficial attraction they have for many, they turn out, on closer analysis, to be flawed. The argument that the law is hypocritical is misconceived and proves too little and too much. Too little in that it fails to dent the clear distinction, well-established in professional ethics and law, between intending and merely foreseeing death, between trying to kill the pain and trying to kill the patient. Too much in that if administering palliative drugs that incidentally shorten life is morally equivalent to VAE, then doing so when the patient is incompetent is equivalent to NVAE. This brings us on to the second argument for decriminalisation, which claims that decriminalising VAE is a discrete step, with no link to NVAE. But as we saw, VAE rests just as much on the doctor's judgement that the patient's autonomous request is justified (that the patient would indeed be 'better off dead') as it does on the patient's autonomous request. There is, therefore, a clear logical link between VAE and NVAE. The third argument, that the current criminal prohibition is ineffective, is unsupported by the evidence. Indeed, what *has* proved ineffective according to the evidence is the Dutch law regulating VAE. Finally, to the argument that a law could distinguish the 'non-vulnerable' from the vulnerable, we shall recall not only the disturbing evidence to the contrary from the Netherlands, but also the judgment of Lord Bingham that such a law 'could not be administered fairly and in a way which would command respect', a judgment that echoes that of many expert bodies, of which the Walton Committee is but one.

As the Walton Committee concluded, the way forward is not to weaken the prohibition on intentional killing. That prohibition is indeed 'the cornerstone of law and of social relationships', which 'protects each one of us impartially, embodying the belief that all are equal'.[68] There is, then, no conflict between the criminal law's prohibition of euthanasia

[67] Director of Public Prosecutions, *Policy for Prosecutors in Respect of Cases of Encouraging or Assisting Suicide* (London: Crown Prosecution Service, February 2010) (see www.cps.gov.uk/publications/prosecution/assisted_suicide_policy.html). For criticisms of the interim guidance, see J. Keown, 'Dangerous Guidance' (2009) 159 *New Law Journal* 1718. To the extent that the final guidance did not incorporate those criticisms, it remains unsatisfactory.

[68] House of Lords Select Committee on Medical Ethics, *Report*, para. 237.

and sound bioethics. In my oral debate on euthanasia with John Griffiths in 2009,[69] I took him to suggest that euthanasia is 'too dangerous' to be left to the criminal law. But it is precisely *because* intentional killing is so dangerous that its prohibition by the criminal law is entirely proper, indeed necessary. The way forward, from which the euthanasia debate is an increasing distraction, is not to weaken the criminal law: it is to improve the quality and availability of palliative and social care, especially to those who are dying, disabled and disadvantaged. The government's End of Life Care Strategy (2008)[70] is a timely and welcome development.

[69] At the *Good, Bad or Indifferent? Medicine and the Criminal Process* conference, the University of Manchester, 4 November 2009.

[70] Department of Health, *End of Life Care Strategy: Promoting High Quality Care for All Adults at the End of Life* (London: Department of Health, 2008).

4 Euthanasia excused: between prohibition and permission

Richard Huxtable

Introduction

A legal system – and, indeed, the citizens to whom its rules apply – is unlikely to get far without taking a stand on the intentional ending of human life. Such behaviour will typically be deemed contrary to the criminal law: in other words, it will be prohibited and, usually, subject to the gravest penalties. But this seemingly fundamental prohibition sometimes comes under attack from those who maintain that there are occasions when causing death should be permitted, such as where the alleged 'victim' desperately wanted to die in order to escape the suffering inflicted from some terminal medical condition. Clashes between scholars like Griffiths (pro-euthanasia[1]) and Keown (anti[2]) reveal some markedly different moral calculations, beyond which there has been relatively little progress in decades.

In this essay I aim to take a different approach to the topic, in an effort to move things forward. Rather than side with the case for or against assisted dying or euthanasia of some form or another, I argue for a compromise position, in which some weight is accorded to each side, but neither has the satisfaction of wholly outweighing the other. The compromise I envisage is one in which talk of what is (or is not) *justified* is replaced with considerations of *excuse*, as articulated in a new offence, called something like 'mercy killing'.[3] In order to get to this conclusion, I begin with a recent case from the English criminal courts, sketch the ethical arguments usually applied to such cases, and then expand on the case for compromise (which I have hitherto only outlined).

Thanks to participants in the 2008 seminar at the University of Manchester, where some of the ideas in this paper were explored (the usual caveat applies); thanks also to the AHRC for funding that seminar, and to the Wellcome Trust for providing research support during the writing of this chapter.

[1] See Griffiths in this volume. [2] See Keown in this volume.

[3] I shall not discuss the different terms (like 'assisted dying', 'mercy killing', etc.) with which this debate is replete: for such discussion, see R. Huxtable, *Euthanasia, Ethics and the Law: From Conflict to Compromise* (London: Routledge-Cavendish, 2007).

Euthanasia: permitted or prohibited?

In that case . . .

At the age of 14, the previously athletic Lynn Gilderdale was diagnosed with myalgic encephalomyelitis (ME), a condition that involves chronic fatigue.[4] As her condition deteriorated, she became confined to her bedroom, unable to swallow or move her legs, and occasionally incapable of recognising her loved ones. She drew up an advance directive refusing resuscitation should it appear indicated, and she also looked into the possibility of travelling to Switzerland, to take up the option of assisted suicide available through the organisation Dignitas.[5] In December 2008, by which time Lynn was 31, she asked her mother, Kay, to help her commit suicide. Mrs Gilderdale initially sought to discourage her daughter, but ultimately she relented, providing her with a dose of morphine, which her daughter administered herself. When Lynn awoke in distress, Mrs Gilderdale gave her sleeping pills and antidepressants, and also injected air into her veins, in an effort to induce an embolism. In the 28 hours it took Lynn to die, her mother investigated suicide techniques both online and in a telephone call to a 'right-to-die' organisation.

On being discovered and charged, Mrs Gilderdale pleaded guilty to assisting an attempted suicide, contrary to the Suicide Act 1961. Not confident that Lynn's death could be attributed to the injection of air (as opposed to the morphine), the prosecutors – under the direction of the Director of Public Prosecutions – decided also to bring a charge of attempted murder. Brown J doubted that this latter decision was 'in the public interest'; why not, he thought, stick with the first charge, 'rather than let this defendant get tangled up in a messy trial for the sake of some legal mumbo-jumbo?'[6]

Undeterred, the prosecutors proceeded to trial, presided over by Bean J. Convicted of the assisting suicide offence, Bean J sentenced Mrs Gilderdale to a twelve-month conditional discharge for aiding and abetting suicide. He noted that it would be 'inexpedient to inflict any punishment' after Mrs Gilderdale's lawyer had argued that the defendant would be unlikely to be prosecuted according to their reading of

[4] S. Bird, 'Kay Gilderdale Cleared of Attempting to Murder Daughter with ME', *The Times*, 25 January 2010; 'Keir Starmer: Decision to Charge Kay Gilderdale for Attempted Murder Was Right', *The Telegraph*, 26 January 2010.

[5] R. Huxtable, 'The Suicide Tourist Trap: Compromise Across Boundaries' (2009) 6 *Journal of Bioethical Inquiry*, 327.

[6] Bird, 'Kay Gilderdale'.

a new policy, which was published shortly after proceedings had begun.[7] It was the jury that rejected the other charge of attempted murder, for which the judge commended them:

> I do not normally comment on the verdicts of juries but in this case their decision, if I may say so, shows that common sense, decency and humanity which makes jury trials so important in a case of this kind.[8]

To prohibit or permit?

Contributors to the debates typically initiated by cases like that of the Gilderdales usually make one of two claims: either someone like Kay Gilderdale should be *permitted* to provide assistance in dying, or they should be *prohibited* from doing so. Each side usually advances two arguments in support of their favoured position.[9]

Advocates of assisted dying favour two claims about the value of life: that it is *self-determined* and that it is *instrumental*. The first of these claims is prevalent in contemporary thinking about the ethics of health care, since it essentially refers to the principle of respect for (patient) autonomy. Autonomy, which derives from the Greek *auto* (self) and *nomos* (governance), directs attention to such issues as the patient's freedom to exercise choice, level of information about the choices available, and mental capacity in making the choice. If the patient does satisfy the relevant criteria, then, the argument runs, the value of life is entirely a question for her – indeed, it would be unjustly paternalistic of anyone to interfere with her view, such as by imposing treatment on her against her will or denying her the right to commit suicide, even with assistance. As Brock explains:

> If self-determination is a fundamental value, then the great variability among people on this question makes it especially important that individuals control the manner, circumstances, and timing of their dying and death.[10]

For some, this argument is sufficient to ground a permissive approach to assisted dying, whether this takes the form of assisted suicide or

[7] Director of Public Prosecutions, *Policy for Prosecutors in Respect of Cases of Encouraging or Assisting Suicide* (London: Crown Prosecution Service, February 2010).

[8] 'Keir Starmer: Decision to Charge'.

[9] See, further, R. Huxtable, 'Calculating the Value of Life?', in D. Dickenson, R. Huxtable and M. Parker (eds.), *The Cambridge Medical Ethics Workbook* (2nd edn) (Cambridge University Press, 2010), p. 2; Huxtable, *Euthanasia*.

[10] D.W. Brock, *Life and Death: Philosophical Essays in Biomedical Essays* (Cambridge University Press, 1993), p. 206.

voluntary euthanasia. For others, as De Haan has explained, the autonomy argument is joined to an argument from mercy or beneficence in order to make the case.[11]

The mercy argument reflects a different perspective on the value of life, in which life has an *instrumental* value. On this view, life essentially derives its value from the uses to which it can be put. A useful or happy life is one that we can describe as having a good quality; conversely, a life of disability, inability and/or suffering might sometimes be described as a poor quality life. Proponents of this viewpoint believe that we are morally obliged to tackle suffering, through treatment and ongoing research. But some defenders of this position go further and argue that it is not necessarily wrong for us to think that some patients' lives are of such desperately poor quality that they would be better off dead.[12] In assisted suicide and voluntary euthanasia, it is the patient who leads with this evaluation; but it need not be her alone, as Doyal proposes:

> If competent patients who are terminally ill should be legally able to choose assisted death because they believe that their lives are no longer worth living then should it not be possible for clinicians in partnership with families to make similar decisions on behalf of those who cannot competently choose for themselves?[13]

Whether all advocates of assisted dying would follow Doyal's proclaimed logic is a moot point. Even if we restrict ourselves to voluntary forms of assisted dying, advocates do differ over the best legal means of expressing the principled positions they advance – for example, over decriminalisation as opposed to legalisation, medical rather than lay involvement, and assisted suicide as distinct from voluntary euthanasia. Where the advocates coincide is in concluding that there is no reason in principle why Lynn Gilderdale should not be helped to die; indeed, they would suggest there are sound principled reasons for allowing this.

Opponents of assisted dying appeal to a different principle and also to more pragmatic concerns about what allowing assisted dying would entail. Fundamentally, they argue, there is a duty to protect life, reflective of its special worth, which at least means that life should never intentionally be brought to a premature end. This account of the *intrinsic* value of life has a long history, and is particularly associated with Judaeo-Christian teaching – hence references to the 'sanctity of life'.

[11] J. De Haan, 'The Ethics of Euthanasia: Advocates' Perspectives' (2002) 16 *Bioethics*, 154.

[12] E.g. P. Singer, *Practical Ethics* (2nd edn) (Cambridge University Press, 1993), p. 191.

[13] L. Doyal, 'Dignity in Dying Should Include the Legalisation of Non-Voluntary Euthanasia' (2006) 1 *Clinical Ethics*, 65.

However, less theistic formulations are available, such as the right to life or the principle of the inviolability of life.[14] Watt outlines the core idea:

> Those who believe in the value of human life have always recognized that there are limits to the duty to preserve the lives of patients ... What the value of life *does* demand is that we recognize this value at all times, always refusing to attack a human life on the grounds that it is not worthwhile. We should not, in other words, buy into the assumption that the lives of some human beings have no value and may be ended on those grounds.[15]

Seen from this perspective, all of society, including patients like Lynn Gilderdale herself, should see every life as worthy of respect and protection. As Watt indicated, this does not commit us to doing everything in all circumstances for all patients: futile or overly burdensome treatments can still be avoided (according to a distinction between 'disproportionate' and 'proportionate' responses), and potentially risky symptom relief can also be undertaken – provided that there is never any intention to shorten life (according to a distinction between foreseen and intended harms, associated with the 'doctrine of double effect').[16]

To this principled objection a practical problem is coupled, which is usually explained in terms of a 'slippery slope': even if we begin with assisted dying on request for terminally ill patients, we shall – in fact or in principle – find ourselves some considerable distance from this starting point.[17] Various calamitous predictions are made and analogies drawn,[18] but critics typically point to the Netherlands, which has the longest permissive history.[19] Reports of up to 1,000 annual instances of 'voluntary' euthanasia, which appear not to have been initiated by an explicit request from the patient, lead opponents like Keown to remark: 'For inhabitants of such a flat country, the Dutch have proved remarkably fast skiers.'[20]

[14] J. Keown, *Euthanasia, Ethics and Public Policy: An Argument Against Legalisation* (Cambridge University Press, 2002).

[15] H. Watt, 'Euthanasia: Unpacking the Debate' (Oxford: The Linacre Centre, 2002) (see www.linacre.org/euunde.html) (original emphasis).

[16] Keown, *Euthanasia*.

[17] Y. Kamisar [1957], 'Euthanasia Legislation: Some Non-Religious Objections', in A. B. Downing (ed.), *Euthanasia and the Right to Death: The Case for Voluntary Euthanasia* (London: Peter Owen, 1969), pp. 85, 106–17.

[18] See Huxtable, *Euthanasia*, pp. 17–21, 27–30.

[19] J. Griffiths, H. Weyers and M. Adams, *Euthanasia and Law in Europe* (Oxford: Hart, 2008).

[20] J. Keown, 'Euthanasia in the Netherlands: Sliding Down the Slippery Slope?', in J. Keown (ed.) *Euthanasia Examined: Ethical, Clinical and Legal Perspectives* (Cambridge University Press, 1997), pp. 261, 289.

These two groups of paired arguments are so familiar that even five decades ago Glanville Williams saw fit to describe the dispute as 'jaded'.[21] Their persistence is surely attributable in part to their resilience, borne of an intuitive plausibility that doggedly survives even the most critical reflection.[22] We should aim to protect and preserve life; we should be careful about the (present and future) scope of the policies we articulate today; but we should also seek to eradicate or at least minimise suffering; and we should also strive to respect people's views on how they would (not) like to be treated.

Yet all four positions are just as susceptible to attack. The evidence for a slippery slope is said to be lacking and appeals to its logical force as an argumentative tool are condemned as fallacious.[23] The sanctity of life position is dismissed as a theological construct, capable only of persuading particular believers, which selectively interprets and applies key terms (like 'intended', 'futile' and 'burdensome') to reach the conclusions it wants. Moreover, some critics think it actually reaches conclusions it expressly does *not* want, since references to 'futile' and 'burdensome' treatments and (merely?) 'foreseen' outcomes rest on concealed judgements about suffering and quality of life, which the sanctity of life approach (apparently) forbids. Such references, say the critics, are more reflective of an instrumental account of the value of life, but this approach also has its problems. What is to count as unacceptable suffering, such that life should be brought to an end? Who is to make the final judgement, particularly in cases where the patient is not competent to do so herself? Once appeals are made to such an ethic, it seems difficult to detect or erect a firm principled barrier that would restrict assisted dying to the terminally (or even the chronically) ill. This problem also features in accounts of the self-determined value of life: it is hard to deny, in principle, *any* autonomous request to die, irrespective of the basis for the request – which, as a Dutch case illustrated, might be (merely?)

21 G. Williams, 'Euthanasia Legislation: A Rejoinder to the Non-Religious Objections', in Downing, *Euthanasia and the Right to Death*, pp. 134, 143. Williams' paper was originally published in 1958.

22 This capacity suggests that the positions are worthy of serious moral consideration, according to the sorts of criteria against which such judgements, principles and theories might be judged: see, e.g., T. Regan, *The Case for Animal Rights* (2nd edn) (Berkeley: University of California Press, 2004), p. 121; L. Pollock, 'Evaluating Moral Theories' (1988) 25 *American Philosophical Quarterly*, 229.

23 E.g. S. Smith, 'Evidence for the Practical Slippery Slope in the Debate on Physician-Assisted Suicide and Euthanasia' (2005) 13 *Medical Law Review*, 17; S. Smith, 'Fallacies of the Logical Slippery Slope in the Debate on Physician-Assisted Suicide and Euthanasia' (2005) 13 *Medical Law Review*, 224.

'life fatigue'.[24] Depending on which account of autonomy is adopted, it seems increasingly hard to rule out – in principle – any demand, howsoever reached, provided that someone is willing (also autonomously) to assist.

So, should we permit or prohibit? Mindful of Williams' comment half a century ago, the conflict is unlikely to abate overnight, but still we need some resolution. There are many routes by which we might reach this destination, and no fewer ways by which we might travel. A party might accede, concede or supersede; agreement, appeasement or capitulation might ensue; and such outcomes might derive from direct discussion and dispute or, whether by desire or dictate, from the intervention of some third party.[25] As the familiarity of the preceding arguments attests, none of these outcomes has quite materialised. There are pockets of seeming agreement – for example, around the need to provide good palliative care, respect consciences and recognise that sometimes treatment can be ceased. But all-out consensus looks highly unlikely and neither side appears to have established their primacy (by some means or another). Given where we are, I suggest that we should move from asking whether we should prohibit or permit, to considering how we might *excuse*; in other words, we should redirect our energies from conflict to *compromise*. This is an argument I have defended elsewhere;[26] in the remainder of this essay I aim to substantiate my position, by paying more attention to the circumstances, conditions and contours of compromise.

Euthanasia: excused

Why (not) compromise?

Etymologically, compromise refers to an agreement that has been reached through mutual concessions;[27] it is, by definition, always sub-optimal.[28] The term can refer to the *procedure* (process) by which the

[24] R. Huxtable and M. Möller, '"Setting a Principled Boundary"? Euthanasia as a Response to "Life Fatigue"' (2007) 21 *Bioethics*, 117.
[25] T. M. Benditt, 'Compromising Interests and Principles', in J. R. Pennock and J. W. Chapman (eds.), *Compromise in Ethics, Law, and Politics* (*NOMOS XXI*) (New York University Press, 1979), p. 26.
[26] Huxtable, *Euthanasia*.
[27] M. Nachi, 'The Morality in/of Compromise: Some Theoretical Reflections' (2004) 43 *Social Science Information*, 291; R. Cohen-Almagor, 'On Compromise and Coercion' (2006) 19 *Ratio Juris*, 434.
[28] T. Van Willigenburg, 'Moral Compromises, Moral Integrity and the Indeterminacy of Value Rankings' (2000) 3 *Ethical Theory and Moral Practice*, 385.

mutual concessions are secured or, alternatively, to the resulting *product* (solution, end-state, outcome or substance). Often, says Kuflik, each party will be involved in both – that is, 'give-and-take by each (process), resulting in some gain and some loss for all (end state)'.[29] Contributions like Kuflik's aside, studies of compromise in economics, international relations and mathematical game theory find few counterparts in ethics and law.[30] This might not be surprising: although positive things can be said,[31] compromise is more often understood in a pejorative sense. Santayana neatly captures the two main objections: 'Compromise is odious to passionate natures because it seems a surrender; and to intellectual natures because it seems a confusion'.[32]

First, then, compromise appears to signal the *surrender* of (moral) integrity. As Benditt says: 'We expect people to stand for something, and we think less of a person who is not willing to espouse any principles, and even worse of one who vacillates'.[33] Martin Benjamin, who leads philosophical thinking about compromise, believes we each invest in our own personal world view:

> A *world view* is a complex, often unarticulated (and perhaps not fully articulable) set of deeply held and highly cherished beliefs about the nature and organization of the universe and one's place in it. Normative as well as descriptive – comprised of interlocking general beliefs about knowledge, reality, and value – a world view so pervades and conditions our everyday thinking that it is largely unnoticed.[34]

A world view, in turn, structures and grounds a way of life. Integrity involves *wholeness* – that is, consistency and coherence between the two – and also *wholeheartedness* – that is, sincere self-direction towards particular commitments, rather than slavishly following one's peers or passions.[35]

Hölm has criticised the argument, which he attributes to me, that we can compromise (or, in Hölm's words, 'agree to disagree') on euthanasia, partly because he anticipates 'significant cognitive dissonance

[29] A. Kuflik, 'Morality and Compromise', in Pennock and Chapman, *Compromise in Ethics*, pp. 38, 40.
[30] See, e.g., M. Benjamin, *Splitting the Difference: Compromise and Integrity in Ethics and Politics* (Lawrence, KS: University Press of Kansas, 1990); Pennock and Chapman, *Compromise in Ethics*; Huxtable, *Euthanasia*.
[31] E.g. Kuflik, 'Morality and Compromise', p. 38.
[32] M. P. Golding, 'The Nature of Compromise: A Preliminary Inquiry', in Pennock and Chapman, *Compromise in Ethics*, pp. 3, 4–5.
[33] Benditt, 'Compromising Interests', p. 31.
[34] M. Benjamin, 'Conflict, Compromise, and Moral Integrity', in C. S. Campbell and B. A. Lustig (eds.), *Duties to Others* (Dordrecht: Kluwer Academic Publishers, 1994), pp. 261, 267–8 (original emphasis).
[35] Van Willigenburg, 'Moral Compromises', 388.

and unease'.[36] His attribution may be accurate, but there are good reasons to resist his objections. As Kuflik reminds us, there is usually more at stake than the disputed issue itself. The promotion or preservation of peace, the avoidance of force, the importance of open dialogue, and the acknowledgement of reasonable moral diversity will all be important features on the landscape of compromise.[37] One's world view (and corresponding ways of being) will admit of such variety; as Benjamin says, 'our identity is constituted in part by a complex constellation of occasionally conflicting values and principles'.[38] Sometimes consistency must give way to this plurality, in keeping with the complicated entirety of the integrated life. 'The resulting ambivalence is part of the price we must pay to avoid the dehumanization of simple consistency in an unavoidably complex situation.'[39] Van Willigenburg adds that compromising on some values can help safeguard others and thus preserve (not imperil) wholeness. But, he says, integrity will also involve *purity* – that is, a commitment to morality's cause, according to which Hitler cannot be dubbed a man of integrity.[40]

Here we encounter the second objection, since people of moral integrity should apparently not *confuse* morality with a 'negotiable commodity'.[41] Duty-bound deontologists, consequence-led utilitarians and character-concerned virtue ethicists will fundamentally disagree about morality's demands, but each will see their account as objective and true. Why should they consider accommodating a mistaken opponent?[42] Disputes about euthanasia reveal such fundamental moral differences that compromise seems inconceivable to commentators like Cohen-Almagor and Tännsjö.[43] Hölm, in particular, has difficulty on those occasions when:

> it is sometimes claimed that what should happen in a liberal society is (merely?) that the two sides in the debate 'agree to disagree' or 'split the difference' (Huxtable 2007); and the Netherlands is taken as an example of this kind of accommodation.[44]

I should initially clarify that I nowhere defend the view that the Netherlands exemplifies a compromise position on euthanasia (for

[36] S. Hölm, 'Euthanasia: Agreeing to Disagree?' (2010) 13 *Medicine Health Care and Philosophy*, 399, 402.
[37] Kuflik, 'Morality and Compromise', p. 51.
[38] M. Benjamin, 'Philosophical Integrity and Policy Development in Bioethics' (1990) 15 *The Journal of Medicine and Philosophy*, 375, 385.
[39] Ibid., 277. [40] Van Willigenburg, 'Moral Compromises', 389. [41] Ibid., 395.
[42] See Benditt, 'Compromising Interests', pp. 31, 34.
[43] Cohen-Almagor, 'On Compromise', 438; T. Tännsjö, 'Why No Compromise Is Possible' (2007) 38 *Metaphilosophy*, 330.
[44] Hölm, 'Euthanasia', 399. 'Huxtable 2007' refers to Huxtable, *Euthanasia.*

which reason Hölm is right to include his citation of my work before the semicolon). Although the Dutch policy of pragmatic tolerance involves something of a social compromise,[45] the Netherlands' approach to euthanasia is distinctively *permissive*.[46]

But is there something in Hölm's resistance to compromise in cases where the disputants' world views are so fundamentally engaged that 'the mere suggestion that there is a possible compromise is seen as grotesque and improper'?[47] Unlike Hölm (and Golding before him[48]), Benditt thinks compromise on questions of principle is conceivable. He points out that it is not the 'principles that are compromised but claims that have been put forward or policies that are favored'.[49] Different weights can be attached to the applicable principles and they might permit of different expressions, which allow '*room* for compromise'.[50]

Moral (or normative) pluralism helps anchor Benditt's position. Isaiah Berlin saw conflicts of values as an unavoidable feature of human coexistence. We should, he said,

> look upon life as affording a plurality of values, equally genuine, equally ultimate, above all equally objective; incapable, therefore, of being ordered in a timeless hierarchy, or judged in terms of some absolute standard.[51]

On this view, moral principles cannot be traced to a single supreme value ('simple-mindedness', says Benjamin[52]) and there is not always necessarily a single correct solution to a moral dilemma.[53]

The difference between the pluralist and the objectivist (monist, universalist or even absolutist) should be evident; the distinctions between the pluralist and the relativist are not always so apparent. Van Willigenburg suggests that, unlike the relativist, the pluralist recognises there is 'something like truth or correctness in normative issues, but this truth is much more complex than we often think'.[54] W. F. May thinks relativists and objectivists have more in common, since both 'subscribe to the view that the validity of a principle depends upon its universality', albeit the former might find an exception to the principle espoused by the latter and see this as reason to reject

45 B. Gordijn, 'Regulating Moral Dissent in an Open Society: The Dutch Experience with Pragmatic Tolerance' (2001) 26 *Journal of Medicine and Philosophy*, 225, 231.

46 See De Haan, 'The Ethics of Euthanasia'.

47 Hölm, 'Euthanasia', 400.

48 Golding, 'The Nature of Compromise'.

49 Benditt, 'Compromising Interests', p. 27.

50 Ibid., p. 34.

51 I. Berlin, *The Crooked Timber of Humanity* (New York: Knopf, 1991), p. 71.

52 Benjamin, 'Conflict, Compromise, and Moral Integrity', p. 272.

53 Van Willigenburg, 'Moral Compromises', 395.

54 Ibid., 396.

the principle. The pluralist, meanwhile, feels the ongoing pull of principles (plural): 'Such is the rough landscape of policy-making in which one may need to compromise, not in the sense of defecting from duty but honoring duties which are multiple.'[55] Sometimes we might be able to rank these duties (on some criteria) – but more often the ranking (and the criteria) will be indeterminate.[56] When we find ourselves at such a crossroads, the moral pluralist would permit – even encourage – us to contemplate compromise.

When to compromise

So, we shall sometimes have cause to compromise – and this need not mean compromising on our causes. But in which circumstances might this appear permissible or even laudable? To my mind, there are six circumstances that drive compromise: prudence, uncertainty, complexity, necessity, coexistence and scarcity.

Prudence simply refers to the possibility of losing out entirely if one is disinclined to compromise.[57] Disenfranchisement is likely to result where, say, defenders of the intrinsic value of life fail to press their claims in a commission on euthanasia.[58] (Of course, on so engaging, these same advocates should appreciate the likelihood of both gaining *and* losing ground.)

Second, empirical and metaphysical *uncertainty* can drive compromise. Factual uncertainty about diagnoses and prognoses, the prospects for scientific progress, and the plausibility of predictions (including those depicting slippery slopes) are unlikely to vanish overnight. Ongoing philosophical disputes about the nature and significance of 'personhood' and the value of life seem similarly intractable, and such difficulties combine in the light of emergent data that suggest that vegetative state patients might retain some capacity.[59] Eradication of uncertainty seems a will-o'-the-wisp, and so the case for compromise gains ground.

These problems, in turn, indicate the existence of a third reason to consider compromise: the presence of moral and conceptual *complexity*.[60] Such complexity is all too apparent in the euthanasia debate, since each of

[55] W. F. May, 'Oral Evidence to President's Council on Bioethics' (2003) (see http://bioethics.georgetown.edu/pcbe/transcripts/oct03/session6.html).

[56] Van Willigenburg, 'Moral Compromises', 396.

[57] Benditt, 'Compromising Interests', p. 34.

[58] See Huxtable, *Euthanasia*, pp. 155–8.

[59] M. M. Monti, A. Vanhaudenhuyse, M. R. Coleman, M. Boly, J. D. Pickard, L. Tshibanda, A. M. Owen and S. Laureys, 'Wilful Modulation of Brain Activity in Disorders of Consciousness' (2010) 362 *The New England Journal of Medicine*, 579–89.

[60] See Benjamin, *Splitting the Difference*, pp. 29–30.

the opposing moral arguments has some merit. In such situations, Benditt encourages us to see our 'opponent as a sincere and intelligent, though mistaken, surveyor of the moral landscape'.[61] Against a pluralistic backdrop, as Van Willigenburg explains, we should acknowledge the ontic and/or epistemic indeterminacy we face: either 'there *is* no superior way of mixing values or *we* are unable to rationally trace that superior mix'.[62] Indeed, open acknowledgement of such complexity can be particularly welcome if it broadens our perspectives, enabling us to see the world through another's eyes.[63]

Fourth, the case for compromise becomes stronger where a decision is *needed*. Achieving or preserving peace might ground this necessity,[64] as might the relative urgency of the situation. As Benjamin notes, 'in a hospital as well as in other practical contexts we cannot always suspend judgment and action'.[65] An issue like euthanasia – where lives are at stake – is scarcely one on which we can afford to be unclear about the (un)acceptable parameters of the judgement and action in question.

Necessity also features in the fifth circumstance in which compromise might be contemplated, since the disputants must *coexist*. Sometimes the direct transaction might be brief, but even that relationship should be safeguarded until the dispute is resolved.[66] In any event, disputes about euthanasia tend to involve participants who cannot simply 'quit the game'[67] – for example, patients, health care professionals and even the health care system at large, whose relationships will be ongoing. 'All of us', says Benjamin, 'whether extreme conservatives, extreme liberals, or some type of moderate, are members of a single nation requiring uniform policy on matters of life and death.'[68]

Finally, and perhaps controversially,[69] *scarcity* of resources can compel compromise. According to Benjamin:

[61] Benditt, 'Compromising Interests', p. 35.
[62] Van Willigenburg, 'Moral Compromises', 400 (original emphasis); see also Tännsjö, 'Why No Compromise'.
[63] Kuflik, 'Morality and Compromise', p. 50.
[64] Ibid., p. 51. [65] Benjamin, *Splitting the Difference*, p. 31.
[66] Contrast Kuflik, 'Morality and Compromise', p. 54.
[67] B. J. Winslow and G. R. Winslow, 'Integrity and Compromise in Nursing Ethics' (1991) 16 *The Journal of Medicine and Philosophy*, 307, 311.
[68] Benjamin, *Splitting the Difference*, p. 164.
[69] This circumstance should not dominate the case for compromise; indeed, we might see as 'shameful' any intrusion of economics into end-of-life decision-making (see M. Epstein, 'Legitimizing the Shameful: End-of-Life Ethics and the Political Economy of Death' (2007) 21 *Bioethics*, 23). This is not quite what is at issue here, however; rather, this circumstance arguably does no more than acknowledge that it will not always be possible to give everyone what they want, i.e. a fully permissive and a fully

We often lack the time, money, energy, and other human and natural resources to satisfy everyone's rights or interests, let alone their wants and desires. And when rights or interests conflict because of scarcity, compromise may seem to be both necessary and appropriate.[70]

The more of these six conditions that are present, the stronger the case is for compromise.[71] I have suggested that the case is certainly strong in relation to euthanasia. The stage for compromise is therefore set: how should it be played out?

How to compromise

Unsurprisingly, given the dearth of philosophical exploration, the ground rules (or necessary and sufficient conditions) for achieving 'integrity-preserving'[72] or 'principled'[73] compromise remain open to question. Side-stepping some of the niceties associated with the perceived relationship between morality *in* compromise and the morality *of* compromise,[74] I suggest that there are four (related) conditions to be met in order to reach a principled compromise: sincerity, legitimacy, reciprocity and veracity.

Van Willigenburg comes closest to my first requirement of moral *sincerity*, when he mentions affording room to 'plausible arguments endorsed by thoughtful and intelligent people', and remarks that 'only answers that we can reflectively endorse will be answers that we can trust'.[75] Here we might recall Benditt's insistence on recognising our opponent as a sincere 'surveyor of the moral landscape'.[76] We should not be willing to endorse – as 'moral' principles – any old bias or prejudice; and there should be little place for such non-reflective thinking when we are contemplating principled compromise. As such, the disputants should come to the negotiating table with considered moral positions, which at least strive to satisfy the sorts of criteria (like clarity, coherence, simplicity and empirical 'fit') with which serious moral thinking should be associated.[77]

prohibitive stance on euthanasia. Understood in this way, this impossibility gives us an additional reason for contemplating splitting the difference between the two positions.

70 Benjamin, *Splitting the Difference*, p. 32; see also Kuflik, 'Morality and Compromise', p. 49.
71 Benjamin, *Splitting the Difference*, p. 32.
72 Ibid.; Winslow and Winslow, 'Integrity'.
73 Cohen-Almagor, 'On Compromise', p. 440.
74 See Kuflik, 'Morality and Compromise'; Nachi, 'The Morality'.
75 Van Willigenburg, 'Moral Compromises', pp. 400–1.
76 Benditt, 'Compromising Interests', p. 35; see also T. V. Smith, 'Compromise: Its Context and Limits' (1942) 53 *Ethics*, 1, 13.
77 E.g. Regan, *The Case for Animal Rights*; Pollock, 'Evaluating Moral Theories'.

Second, building on Benditt, each disputant should see the other as a *legitimate* moral contender. Negotiation is the art of compromise;[78] this requires effective communication; and communication, says Golding,

> presupposes a common language [and] a commonality or, more exactly, a community. The compromise process is a conscious process in which there is a degree *of moral acknowledgement* of the other party. The other party is accorded some degree of *moral legitimacy*, and so are some of his interests. This is the special quality that characterises the cooperative attitude in the compromise process.[79]

Moral differences should not be inflated.[80] Among likely areas of agreement will be an interest in peaceful resolution of the dispute.[81] The legitimacy criterion asks that, in seeking such resolution, each party recognises the other.

Such recognition feeds into the third condition, which might variously (and successively) be described in terms of tolerance, mutual respect, cooperation, a 'democratic spirit' or, my preferred label, *reciprocity*. We start with tolerance, which is needed because – quite apart from any pockets of convergence – there is obviously some divergence. Compromise, says Benditt, 'is possible only if one recognizes a plurality of interests or some principle of tolerance'.[82] Accompanying this recognition should be mutual respect: 'Compromise without reciprocal respect is likely to be mere capitulation', say Winslow and Winslow.[83] This respect entails a cooperative attitude, in which each party must 'attempt to *persuade*, rather than intimidate, the other side to accept its offers', in the manner of a debate.[84] This attitude obviously affects the compromise process; however, it also constrains the *substance* of the compromise that might be reached. For Golding, 'bargaining in "good faith"' means 'responding to a concession with a concession'.[85] Van Willigenburg also thinks that mutuality 'will favour evenly balanced compromises (splitting the difference)'; a 'willingness to strive for some balance in the sort and amount of concessions by different parties'.[86] But balance need not require absolute equality, if such equality would mean an imbalance of power or overlooking a legitimately vested

[78] See Cohen-Almagor, 'On Compromise'.
[79] Golding, 'The Nature of Compromise', p. 16 (original emphasis); see similarly Benditt, 'Compromising Interests', p. 29.
[80] Winslow and Winslow, 'Integrity', 316.
[81] Golding, 'The Nature of Compromise', p. 14; Nachi, 'The Morality', 296–7.
[82] Benditt, 'Compromising Interests', p. 30; see also Winslow and Winslow, 'Integrity'.
[83] Winslow and Winslow, 'Integrity', 317; cf Cohen-Almagor, 'On Compromise'.
[84] Golding, 'The Nature of Compromise', pp. 17–18.
[85] Ibid., p. 17.
[86] Van Willigenburg, 'Moral Compromises', 400, 401.

interest.[87] Sometimes, mutual respect will require recognition that one party has more to lose than the other, and, accordingly, an appropriate adjustment in the compromise that is sought and secured.

Finally, principled compromise requires *veracity*. Implicit in the other criteria, this merits mention as a criterion in its own right, if only to help ward against parties advancing spurious claims in order to secure 'compromises' that in truth reflect the positions they would actually defend.[88] As Golding insists, we need each party to 'be trusted to mean what he says and to mean it sincerely – one must tell the truth and sincerely intend to do what one says one intends to do'.[89] O'Neill reminds us that we only need trust where 'we lack certainty about others' future action';[90] positions of moral conflict surely qualify, so an insistence on veracity here should safeguard the negotiations.[91]

Taken together, these four conditions seem to convey the 'internal morality' of principled compromising – that is, the conditions that must be satisfied if an integrity-preserving compromise, premised on acknowledgement of moral pluralism, is to be attained.[92]

How to compromise on euthanasia

There seems to be ample cause to contemplate compromise as an answer to the euthanasia question. We have some criteria for achieving this. How, then, should we proceed? In keeping with the arguments developed by Benjamin, one way might be to focus on *process*: we could convene a committee, tasked with splitting the difference between the different perspectives on assisted dying. However, given the stability of the arguments on each side, I think we can go further, by describing the *substance* of what such a policy could look like.

In order to achieve this, we must first observe the ground rules of compromise – that is, we should recognise the legitimacy of the opposing claims, engage with the reflective (sincere and true) accounts thereof, and ensure that concessions on one side are (reciprocally) met with concessions on the other. In doing all this, I believe we shall inevitably re-engage with the ambivalence that Robert Burt detects at the heart of these debates: ambivalence about death and its rights and wrongs. Critics of the inviolability of life have long contended that it is only

87 Ibid., 402. 88 Benditt, 'Compromising Interests', p. 36.
89 Golding, 'The Nature of Compromise', p. 18.
90 O. O'Neill, *Autonomy and Trust in Bioethics* (Cambridge University Press, 2002), p. 13.
91 Compare Cohen-Almagor, 'On Compromise', who links trustworthiness to negotiation and communication skills.
92 Whether they are sufficient remains open to question.

'rational' to dismantle this ethic; yet our resistance to embracing death persists, due, says Burt, to 'some lurking, ineradicable sense of its wrongfulness'.[93] Traditional distinctions between acting and omitting and between intended and foreseen outcomes persevere, since they help express our ambivalence, as the rules they birth are 'serving simultaneously as permission and a warning sign about dealing with death'.[94] By clinging to such allegedly illogical rules, we thus remain aware of the acute moral discomfort that comes with cases like that of the Gilderdales.

The moral language capable of describing this ambivalence already exists, although it rarely (overtly) features in discussions about euthanasia: it is the language of *excuse*. Disputants like Griffiths and Keown are more interested in the presence or absence of a *justification* for euthanasia. According to Austin, those who plead justification for an action 'accept responsibility but deny that it was bad', while those who plead an excuse 'admit that it was bad but don't accept full, or even any responsibility'.[95] Also writing in the 1950s, Ramsey recognised that the idea of an excuse (or partial defence to a crime) communicates 'a sense of the accused's tragic involvement in some unspecifiable degree of guilt', and simultaneously, as Kalven added, conveys 'compassion for the actor' and 'compassion for the suffering of the subject'.[96]

The simultaneous recognition of wrong- and right-doing that is at the heart of an excuse does, in truth, already (covertly) feature in the criminal law's dealings with assisted dying. This has long been the case in England and Wales: stern pronouncements that assisted dying is contrary to the law of homicide (and specifically murder) have scarcely translated into practice, as mechanisms have been employed to ensure lenient disposals (e.g. through the virtually routine acceptance of a plea of diminished responsibility) or even the absence of any criminal proceedings (particularly, it seems, where medical professionals are concerned).[97] Kalven proposed 'trusting for awhile yet to the imperfect but elastic equity in the administration of the law as written',[98] an idea that found recent support in McCall Smith's preference for a 'zone

[93] R. A. Burt, 'The End of Autonomy' in *Improving End of Life Care: Why Has It Been So Difficult?* (2005) 35 *Hastings Center Report Special Report*, s9–s13, s11.
[94] Ibid., s11–12.
[95] J. L. Austin, 'A Plea for Excuses' (1956–57) 57 *Proceedings of the Aristotelian Society*, 1.
[96] P. Ramsey, 'Freedom and Responsibility in Medical and Sex Ethics: A Protestant View' (1956) 31 *New York University Law Review*, 1189, 1201; H. Kalven, 'A Special Corner of Civil Liberties: A Legal View I' (1956) 31 *New York University Law Review*, 1223, 1235.
[97] See Huxtable, *Euthanasia*, Chapters 2, 3 and 4.
[98] Kalven, 'A Special Corner', 1237; see also Ramsey, 'Freedom and Responsibility', 1201.

of indeterminate application' for the humane disposal of the mercy killer.[99] There is something to commend in this flexibility, but it brings with it a considerable measure of uncertainty, unpredictability and unfairness, which signals that important legal values are not being honoured.[100] There are too many cases in which apparently similar cases have resulted in markedly different results: indeed, around the same time that Kay Gilderdale was prosecuted for two crimes, but ultimately convicted of only one and sentenced to a conditional discharge, another mother, Francis Inglis, was sentenced to life imprisonment and ordered to serve no less than nine years in prison, for murdering her brain-damaged son.[101] There are undeniable differences between the cases, not least concerning the capacity to consent of each of the 'subjects' (to use Kalven's word), but whether the same system should permit of such radically different disposals should prompt us to pause.

On pausing, we should consider whether recent developments in the prosecution of English cases pave the way for a more consistent compromise approach. Following Debbie Purdy's successful petition of the courts in 2009,[102] the Director of Public Prosecutions (DPP) issued guidance for prosecutors, intended to supplement the *Code for Crown Prosecutors*, which is designed to help them decide which allegations should proceed to court and which should not.[103] Notably, the features we would expect to be present in a paradigmatic case of assisted dying, such as a clear, voluntary wish and an assistant motivated by compassion, are cited as 'public interest' grounds for not prosecuting. But other features that proponents would want to see in a system of legalised or decriminalised assisted dying, such as assistance by a health care professional, appear to indicate in favour of prosecution. Furthermore, as the DPP himself made clear in a press release, this is 'not a tick-box exercise';[104] each case will be considered on its merits and the policy only applies to allegations of assisted suicide.[105] As such, one might doubt whether this policy achieves much, in terms of the compromise

[99] A. McCall Smith, 'Euthanasia: The Strengths of the Middle Ground' (1999) 7 *Medical Law Review*, 194, 197.

[100] R. Brownsword, '"Good Faith in Contracts" Revisited' (1996) 49 *Current Legal Problems*, 111; Huxtable, *Euthanasia*, pp. 144–5.

[101] J. F. Rojas, '"Mercy killing" mother is jailed for life', *The Independent*, 20 January 2010.

[102] *R (on the application of Purdy)* v. *DPP* [2009] UKHL 45.

[103] Director of Public Prosecutions, *Policy for Prosecutors*.

[104] Crown Prosecution Service, 'DPP Publishes Assisted Suicide Policy', 25 February 2010 (see www.cps.gov.uk/news/press_releases/109_10).

[105] I.e. 'encouraging or assisting suicide', contrary to the Suicide Act 1961, as amended by the Coroners and Justice Act 2009.

we are seeking: the seeds are sown for a more consistent clemency, but unpredictability and selectivity surely remain.

I think we can go further: falling between prohibition and permission, we should explicitly mark out assisted dying as a particular – compassionate – type of killing, chargeable as a distinct offence and also available as a partial defence to other homicide charges. The idea of a new offence, called something like 'mercy killing', is not itself new; various commissions and committees have considered the idea,[106] and it already features in other legal systems.[107] Indeed, the offence enshrined in our own Suicide Act provides something of a model, since it marks out assisted suicide as a distinct offence, susceptible to a milder sentence (up to fourteen years) than might be expected for murder. A similar approach seems warranted here, and research into public attitudes suggests 'that a community-based disposal, with the emphasis on counselling for the killer, would be appropriate'.[108] Of course, it might also be appropriate to keep the upper limit of life imprisonment, not least to enable sterner sentencing in more questionable cases (e.g. when motives appear to be mixed). In more routine cases the approach would be different, but this raises the obvious question: what would count as a routine case in such a system?

Qualifying criteria for such an offence have occasionally been proposed.[109] In keeping with definitions of euthanasia,[110] the offence should essentially encompass the defendant who intends to kill the victim, who does indeed die (although we should be open to the possibility of charging attempts), from the motive that they wished to bring an end to the patient's pain, suffering and/or distress. Claims that the criminal law cannot judge motive will ring hollow, not least in the wake of the DPP's guidance, which explicitly refers to compassion. Qualifying cases should be prosecuted consistently, irrespective of the professional status of the defendant, provided that the requisite motive and intent are present. This should probably also mean that appropriate omissions should fall within the remit of the offence.

[106] E.g. Royal Commission, *Royal Commission on Capital Punishment 1949–1953: Report* (London: HMSO, 1953) pp. 63–4; Criminal Law Revision Committee, *Working Paper on Offences Against the Person* (London: HMSO, 1976), pp. 31–4; House of Lords Select Committee, *Report of the Select Committee on Murder and Life Imprisonment* (London: HMSO, HL 78 I–III, 1989); Law Commission, *Murder, Manslaughter and Infanticide. Project 6 of the Ninth Programme of Law Reform: Homicide* (London: Law Commission, Law Com No. 304, 2006).

[107] M. F. A. Otlowski, *Voluntary Euthanasia and the Common Law* (Oxford: Clarendon Press, 1997), pp. 460–1.

[108] Law Commission, *Murder, Manslaughter and Infanticide*, p. 149.

[109] House of Lords Select Committee, *Report of the Select Committee*, pp. 669–71.

[110] See Huxtable, *Euthanasia*, pp. 4–9.

There is obviously more work to be done in substantiating these criteria. One of the thornier dilemmas concerns the prospect of involuntary euthanasia, in which a competent victim is not consulted about, or does not consent to, being killed. It is difficult to detect anyone who would (ethically) defend such a practice, although, remarkably, the judges to date do appear to be just as lenient in these cases as in those where the victim wanted to die.[111] Although it would benefit from further exploration, it is possible that even this sort of killing should fall within the new offence, although, in recognition of its dubiousness, a minimum sentence of imprisonment might be appropriate in such cases.[112]

Leaving aside that complication, the central case should be clear: in recognition of proponents' arguments, assisted dying can be treated as a special case in which greater leniency can be expected; but, in acknowledgement of opponents' arguments, it is still contrary to the criminal law. Proponents will think this does not go far enough. Opponents will think it goes too far, although, as Sanders commented in 1969, 'perhaps the anxiety and discomfort of going through a criminal trial is both a sufficient deterrent to others and an adequate display of public censure'.[113] The likelihood of these very responses suggests we have indeed found the middle ground.[114]

Conclusion

Falling somewhere between the 'all' that proponents of assisted dying demand and the 'nothing' claimed by its opponents, I have defended a compromise approach to the phenomenon, which is readily amenable to articulation in the law. Rather than invest our increasingly depleted energies in a very familiar to-and-fro, I think there is merit in further exploring the terrain of the middle ground. In an effort to advance this exploration, I dwelt at some length on the circumstances and conditions of compromise. There is undoubtedly more to be said here, particularly about the *limits* of compromise. When might we say that compromise should not be attempted? There are suggestions (in what little literature there is) that the prospects for compromise are exhausted when the attempt is doomed (e.g. when dealing with a despot, who does not

[111] Ibid., p. 49. [112] See, further, ibid., pp. 170–1.

[113] J. Sanders, 'Euthanasia: None Dare Call It Murder' (1969) 60 *The Journal of Criminal Law, Criminology, and Police Science*, 351, 358.

[114] D.W. Meyers, *The Human Body and the Law: A Medico-Legal Study* (Edinburgh University Press, 1970), p. 155.

take the demands of morality seriously), or when to do so would mean an end to one's moral agency (e.g. by entering one into slavery).[115] Whether these are sufficient or necessary boundaries is a debate for another day. For now, I hope to have mounted a plausible case for a different, morally sensitive approach to assisted dying, in which notions like excuse supplant our usual preoccupation with the presence or absence of justification.

[115] Kuflik, 'Morality and Compromise', pp. 46, 52; see also Smith, 'Compromise', 9.

Part II

Freedom and autonomy: when consent is not enough

5 Body Integrity Identity Disorder: a problem of perception?

Robert C. Smith

Introduction

Body Integrity Identity Disorder (BIID) is the desire for the amputation of a healthy limb. In my role as a Scottish National Health Service (NHS) general surgeon, I performed two of these amputations in 1997 and 1999. I was initially approached by the first patient, who knew of my interest in amputation surgery through my specialty interest, at that time, of vascular surgery (vascular reconstruction is not always successful in salvaging a critically ischaemic limb). My initial response was that this was a crazy request, but, after discussion with the patient, I became more convinced of the existence of a genuine "need" to be an amputee. The desire seemed to be based not just on enhancing appearance, but on a true manifestation of "core identity". A potent stimulus to my interest was the comment of his wife, who stated, 'I wish someone would just amputate his leg so that we can then have a normal life.'

However, such amputations are clearly highly controversial and attract a great deal of press and public interest.[1] In this essay, I examine the condition of BIID in order to explore a number of fundamental ethical and legal questions that are raised by this condition and its possible treatment. The ethical issues in these cases centre on the issue of patient autonomy. I shall consider whether respect for patient autonomy justifies surgery being provided for those with BIID, or whether other issues, such as harm prevention, trump respect for autonomy. In the light of this ethical discussion, I shall consider what the role of the law is and should be in such cases. Can there be a duty on the NHS to provide amputation in these circumstances? And what of the criminal law? For many people, what I did, as a surgeon, in performing such amputations was not a form of medical treatment, but

[1] BBC News, 'Surgeon Defends Amputations' (see http://news.bbc.co.uk/1/hi/scotland/625680.stm) and BBC News, '"No regrets" for healthy limb amputee' (see http://news.bbc.co.uk/1/hi/scotland/632856.stm).

grievous bodily harm. The crucial question is: should surgeons be protected from criminal sanctions in cases of BIID amputation?

What is BIID?

The desire for amputation of a healthy limb has been recognised since a case report in 1977 by Money *et al.*, who coined the term 'apotemnophilia'.[2] This, however, implied that the condition was a paraphilic or fetish condition. Whilst there is undoubtedly a fetish related to the desire to be an amputee, not all of those desiring to be an amputee are paraphilic. First suggested that the condition should be called 'Body Integrity Identity Disorder' (BIID), thus removing the condition from the classification of paraphilia.[3] This is a condition characterised by a mismatch between an individual's actual body image and their inner perceived body image. It takes the form of a "need" to be "disabled" in a variety of different ways. The commonest form is the "need" to be an amputee, but in other individuals, the condition may manifest itself as a "need" to be paralysed, hearing impaired, visually impaired, incontinent or to have to wear braces or casts. All these phenomena seem to be contrary to the normal human desire to be well and physically perfect.

Features of BIID (amputee form)

BIID usually commences at an early age (as young as four or five years old) and is of lifelong duration. It most often manifests as a desire for the amputation of one leg, but some individuals may desire the amputation of more than one limb. The desire is completely inexplicable to the involved individual and most are ashamed about their feelings. Such feelings of embarrassment and shame often prevent the individual from discussing their feelings with their therapist. The condition causes considerable distress, is all-consuming at times, but waxes and wanes over the course of a lifetime. Some individuals may go to the extent of injuring themselves by lying on railway lines or using a gun in order to achieve the desired amputation. A method of using dry ice (solid carbon dioxide) to achieve irreversible cold injury is described on the Internet.[4]

[2] J. Money, R. Jobaris and G. Furth, 'Apotemnophilia: Two Cases of Self-Demand Amputation as a Sexual Preference' (1977) 13 *Journal of Sexual Research* 115.

[3] M. First, 'Desire for Amputation of a Limb: Paraphilia, Psychosis or a New Type of Identity Disorder' (2005) 34 *Psychological Medicine* 1.

[4] See, for instance, http://transabled.org/stories/my-failed-attempts-and-what-followed.htm.

The condition often commences when an individual with the particular disability is recognised by the individual with BIID as having the ideal body image. It seems to awaken an internal identity that had previously been unrecognised. Subsequent contact with individuals with the particular disability engender feelings of extreme jealousy and the desire to achieve that image. BIID can affect anyone, but individuals with the condition are more commonly of Caucasian origin, male and in intellectually higher levels of society.

One condition that is sometimes confused with BIID is Body Dysmorphic Disorder. This is a psychiatric condition in which the individual is obsessively concerned about and preoccupied by an imagined or minor defect in their physical appearance.[5] They frequently request surgical intervention, but, if operated on, are frequently dissatisfied and request further surgery. In contrast, the BIID individual sees nothing wrong with the involved body part, but sees it simply as not being consistent with their inner body image.

A condition that has the greatest similarity to BIID is Gender Identity Disorder (GID). In this condition, the individual has been assigned a gender, usually on the basis of their sex at birth, but identifies as belonging to another gender and feels significant discomfort or an inability to deal with this situation. Some go so far as to attempt to amputate their genitalia.[6] The aetiology is obscure, but Zhou *et al.* did find some female features in the brains of male-to-female GID sufferers.[7] Lawrence has compared BIID to GID and showed striking similarities between the two conditions.[8] Both commence in childhood, cause great emotional distress, have a significant effect on the individual's working life, are unresponsive to conventional psychiatric treatment and (anecdotally in BIID[9]) both respond well to surgery.

The transsexual (with GID) wishes to be seen and recognised as of the opposite sex. The individual with BIID wants to be seen and recognised as a disabled person. They may pretend to be so in public and, if successful in achieving amputation, they may not wear prostheses so that they are obviously an amputee. This is quite different from the

[5] D. Veale, K. Gournay, W. Dryden, F. Shah, R. Willson and J. Walburn, 'Body Dysmorphic Disorder: A Survey of 50 Cases' (1996) 169 *British Journal of Psychiatry* 196.

[6] P. Cohen-Kettenis and L. Gooren, 'Transsexualism: A Review of Aetiology, Diagnosis and Treatment' (1999) 46 *Journal of Psychosomatic Research* 315.

[7] J. Zhou, M. Hofman, L. Gooren and D. Swaab, 'A Sex Difference in the Human Brain and its Relationship to Human Sexuality' (1995) 378 *Nature* 68.

[8] A. Lawrence, 'Clinical and Theoretical Parallels between Desire for Limb Amputation and Gender Identity Disorder' (2006) 35 *Archives of Sexual Behaviour* 263.

[9] T. Bayne and N. Levy, 'Amputees by Choice: Body Integrity Identity Disorder and the Ethics of Amputation' (2005) 22(1) *Journal of Applied Philosophy* 75.

non-BIID amputee, who usually wishes to be cosmetically restored. The image someone with BIID wishes to present is that which coincides with their internal body image. This is a fundamental part of their self-image and is immutable.

Aetiology

The aetiology of the condition is obscure, but there have been a number of theories:

- an attempt to gain attention from cold and unloving parents; being disabled is seen as a very positive state, likely to gain the required love and attention;
- the sight of a disabled person in childhood imprints on the psyche as an ideal body image;
- the desire is a manifestation of an unresolved internal conflict;
- the condition is a neurophysiological condition caused by an abnormality of the "hard wiring" of the brain present since birth.[10]

The psychiatric and psychological assessments of those with BIID have failed to identify any recognisable psychiatric condition other than anxiety and depression as a result of the underlying problem. Various forms of treatment, including psychotherapy,[11] cognitive behaviour therapy,[12] hypnosis, psychoanalysis and pharmacological therapy,[13] have all failed to provide lasting relief. The achievement of the desired image by surgical means seems, from anecdotal reports, to be the only intervention that results in permanent symptom relief. I personally know of nine BIID patients who have achieved amputation and all are happy with their transformation.

However, if the condition is neuropsychological, and not psychiatric, conventional treatments for mental illness may be unsuccessful. In those situations, some form of physical treatment may be the only treatment to be effective. BIID presents with similar symptoms and signs to the rare and unusual condition called misoplegia.[14] In this condition, a

[10] P. McGeoch, V. S. Ramachandran and D. Brang, 'Apotemnophilia: A Neurological Disorder' (2008) 19 *Neuroreport* 1305.

[11] J. Kasten, 'Body Integrity Identity Disorder (BIID): Befragung von Betroffenen und Erklärungsansätze' (2009) 77 *Fortschritte in Neurologie und Psychiatrie* 16.

[12] A. W. Braam, S. Visser, D. C. Cath and W. J. Hoogendijk, 'Investigation of the Syndrome of Apotemnophilia and Course of a Cognitive Behavioural Therapy' (2006) 39 *Psychopathology* 32.

[13] T. N. Wise and R. C. Kalyanam, 'Amputee Fetishism and Genital Mutilation: Case Report and Literature Review' (2000) 26 *Journal of Sex & Marital Therapy* 339.

[14] J. M. Pearce, 'Misoplegia' (2007) 57(1) *European Journal of Neurology* 62.

lesion in the right parietal cortex results in the patient viewing their left leg as "alien", and they may actually attempt to harm the limb. It is thus of interest that the commonest variety of BIID is the desire to have an above-knee amputation of the left leg. However, although a right parietal anomaly may explain the desire for a left leg amputation, it does not explain the need in some individuals for amputation of right-sided limbs or of more than one limb.

Ramachandran and McGeoch have suggested that stimulation of the cerebral cortex by vestibular caloric stimulation could help BIID symptoms in a similar way to the effect on misoplegia.[15] McGeoch *et al.* have also demonstrated neurological anomalies in the leg that BIID patients want amputated.[16] This would tend to support the hypothesis that BIID is a neuropsychological condition rather than a psychiatric one.

Is it ever ethically and legally acceptable to amputate the healthy limb of someone with BIID?

The duty not to harm and the principle of respect for individual autonomy

One of the doctor's primary ethical obligations is to do no harm to the patient, and the reaction of most doctors to the concept of amputating a healthy limb is that this is totally contrary to this tenet. Therefore, there are those who argue that surgeons who accede to the request for amputation are harming their patient and acting unethically. To the general public, amputation seems to be a mutilating and disabling procedure and thus not ethically acceptable for a normal limb.

However, the subjective perception of the individual with BIID is that amputation is the answer to their problem and will transform their life for the better. They regard amputation as enhancing or enabling rather than disabling. It also prevents them from embarking on attempts at self-injury, which may have fatal consequences. In this respect, it is a harm-avoidance strategy rather than a harmful procedure.

[15] V. Ramachandran and P. McGeoch, 'Can Vestibular Caloric Stimulation Be Used to Treat Apotemnophilia?' (2006) 69 *Medical Hypotheses* 250.

[16] P. McGeoch, D. Brang and V. S. Ramachandran, 'A New Right Parietal Lobe Syndrome?', in A. Stirn, A. Thiel and S. Oddo (eds.), *Body Integrity Identity Disorder: Psychological, Neurobiological, Ethical and Legal Aspects* (Lengerich/Berlin: Pabst, 2009), p. 225.

As well as being under a duty not to harm, doctors also have a strong ethical and legal duty to respect the individual autonomy of their patients. This so-called principle of respect for individual autonomy is arguably the most central ethical principle in modern medical ethics. It requires health care professionals to respect the authentic choices of their patients about their medical treatment. This means that if the decision regarding treatment is made by a patient *who has the requisite mental capacity* to be able to make a decision about their treatment and the relevant information and freedom to make this choice, the decision must be respected by health care professionals, even if it seems foolish or unwise to others. The reason why respect for individual autonomy is seen as so fundamental, ethically (and legally) speaking, is that it enables individuals to have control over their own lives. Indeed, this ability to be in control of our own lives is seen to be more important than being protected from the possible harmful consequences of our choices. It is this ethical principle of respect for individual autonomy that underpins the law on consent to treatment in most jurisdictions, making it a battery to treat without or against a competent patient's consent to this treatment.[17] Thus, so long as a patient demonstrates that they have the requisite mental capacity, are not unduly influenced and have and understand the relevant information, their choices to refuse even life-saving treatment must be respected.[18]

However, if it can be shown that a patient does not have the requisite mental capacity to provide a valid consent to treatment or refusal of treatment, then the principle of respect for individual autonomy no longer requires that we respect their decisions.[19] This ethical principle only requires that health care professionals respect *authentic* decisions – that is, decisions that we believe the patient has the mental capacity, information, freedom, and so on, to make. Those who do not have the requisite capacity to make decisions about their treatment – for example, young children, unconscious individuals, individuals whose capacity to process information has been compromised by a psychiatric illness – should be treated in what is considered to be their best interests, in order to protect them from harm.[20]

Thus, the ethical and legal issues here seem to focus primarily around the mental capacity of patients to make decisions about amputation. It

[17] *F* v. *West Berkshire Health Authority* [1989] 2 All ER 545.
[18] See, for example, *B* v. *NHS Hospital Trust* (2002) 65 BMLR 149.
[19] See the Mental Capacity Act 2005.
[20] In the case of those aged over sixteen who lack capacity, see the Mental Capacity Act 2005, sections 1(5) and 4.

seems that if a BIID patient who makes a decision to request amputation has the mental capacity and understanding to make this decision, then this decision should be respected. Amputation is the treatment the patient desires, and failure to provide the procedure breaches the ethical principle of individual autonomy, as long as the patient is competent to make the decision. If, on the other hand, patients with BIID are considered to be mentally incompetent, they must be protected from their incompetent decision to amputate.

The issue of capacity to give informed consent is explored further in the next section.

Consent

So much hinges on the assessment of a patient's competence to give consent for the procedure. This depends to a certain extent on whether the condition is of psychiatric or neuropsychological origin, a matter I shall return to below.

The General Medical Council (GMC) guidelines on the issue of consent are more concerned with the refusal of a procedure than the request for a procedure. The GMC guidelines state:

64. You must work on the presumption that every adult patient has the capacity to make decisions about their care, and to decide whether to agree to, or refuse, an examination, investigation or treatment. You must only regard a patient as lacking capacity once it is clear that, having been given all appropriate help and support, they cannot understand, retain, use or weigh up the information needed to make that decision, or communicate their wishes.
65. *You must not assume that a patient lacks capacity to make a decision solely* because of their age, disability, appearance, behaviour, *medical condition (including mental illness)*, their beliefs, their apparent inability to communicate, or *the fact that they make a decision that you disagree with.*[21]

Individuals with BIID are, in general, well-educated Caucasians,[22] who have widely researched the subject of amputation, and many of whom will have "rehearsed" (pretended to be an amputee) in private and in public. They have read about the techniques of surgery, rehabilitation

[21] General Medical Council, *Consent: Patients and Doctors Making Decisions Together* (London: General Medical Council, 2008) (see www.gmc-uk.org/static/documents/content/Consent_0510.pdf) (emphasis added).
[22] First, 'Desire for Amputation'.

and prosthetic use. Clearly, no one can fully appreciate what it is like to be an amputee; and as one ages and develops limited cardio-respiratory reserve (walking with an above-knee artificial leg requires more than twice the energy output of walking with two normal legs), mobility may become more difficult. This may lead to dissatisfaction in later years for those who achieve amputation.

As amputation is an irreversible treatment, it is incumbent on the doctor to attempt to treat the incompetent patient by non-surgical methods to enable them to at least live with the condition, even if treatment does not "cure" the condition. Treatment of the underlying illness causing the incompetence may resolve the symptomatology. However, if the condition is neuropsychological, and not psychiatric, conventional treatments for mental illness may be unsuccessful. In those situations, some form of physical treatment may be the only treatment to be effective.

As we have seen, whether the desire to have an amputation is evidence of a psychiatric illness is open to debate. If BIID is a neuropsychological illness (i.e. a structural problem in the central nervous system), the ethics applicable to psychiatric illness may not apply. Notwithstanding this, Müller argues that even if BIID is a neuropsychological condition, patients lack insight into their illness and therefore must be protected from their desire for amputation.[23]

However, my own experience of these individuals is that they are the most well-informed patients I have ever treated. As I have already explained, they have no identifiable psychiatric condition other than the reactive problems due to their unresolved internal conflict. They are fully aware of the consequences of their demand, and those who have succeeded in achieving their aim appear to have lasting relief of symptoms and lead useful, happy and productive lives.[24] In my experience, although BIID sufferers admit that their desire appears to be "nuts" to non-sufferers, they are all fully aware of the consequences of their decisions; they have no identifiable cognitive impairment and present as entirely competent individuals. Therefore, there appear to be no grounds for assessing them as being incapable of giving informed consent. If these individuals seem to be perfectly competent to make the decision to amputate, and we know from those who have achieved amputation that surgery produces complete relief of symptoms, are we

[23] S. Müller, 'Body Integrity Identity Disorder (BIID): Is the Amputation of Healthy Limbs Ethically Justified?' (2009) 9 *American Journal of Bioethics* 36.

[24] G. Furth and R. Smith, *Amputee Identity Disorder: Information, Questions, Answers and Recommendations about Self-Demand Amputation* (Bloomington, IN: Authorhouse, 2000); Bayne and Levy, 'Amputees by Choice'.

ethically justified in withholding operation when at present we have no other form of therapy that provides long-term relief?

Even if the issue of capacity is solved, however, there remain other ethical and legal issues that need consideration.

Justice

Amputation seems entirely just for the individual with BIID, but perhaps not so for the community at large. The general public regard an amputee as someone who is disabled and therefore, by definition, dependent. Undoubtedly, an amputee will have some problems of mobility and self-care, and these may be perceived as creating problems and expense for the community. For example, they may require adapted housing, prosthetic fitting, income support or help with transport. There is also the cost of surgery.

However, individuals with BIID wish to be seen as amputees and also wish to rise to the challenge of apparent disability, which they contrarily regard as enabling. Anecdotal reports suggest that those who achieve amputation become more productive, happy and contributing members of the community. They no longer consume psychiatric resources, as their reactive psychiatric problems resolve.

It is also cheaper and safer to carry out an elective amputation than to deal with an emergency situation of serious acute limb injury in someone who has attempted self-injury. The two patients I operated on were discharged from hospital after their elective amputations within three days. No patient having a traumatic amputation would recover that quickly, and, to that, it is necessary to add the costs of the recovery from the scene of "accident" by the emergency services and the initial emergency treatment.

Legal issues

There appear to be two major areas of legal concern in the proposal that BIID sufferers should be offered amputation: the question of whether there can be legal redress if the individual is refused what they regard as necessary treatment and whether legal action can be taken against the surgeon for performing an amputation.

Legal redress if the individual is refused what they regard as necessary treatment In public law, the individual refused treatment could pursue an action for judicial review against the local Primary Care Trust or claim a breach of the Secretary of State's duties under the National Health Service Act 2006,[25] but in both scenarios, the individual is likely to be unsuccessful. There are a number of reasons for this:

i. The state is under no legal obligation to provide medical care to its citizens.[26] Such an obligation may only be enforced through the ballot box.
ii. Although the Acts of Parliament related to the National Health Service create all sorts of obligations, these can rarely be related to individual patient treatment.[27] Only if the patient can establish that a decision to withhold treatment is "*Wednesbury*[28] unreasonable" (so unreasonable that no competent and conscientious decision maker could have arrived at it), or irrational,[29] could they have redress. Essentially, to bring a successful action for judicial review, a patient would need to identify some obvious flaw in the decision-making process that led to the refusal to provide treatment.[30]
iii. The Human Rights Act 1998 does impose some legally enforceable obligations upon public authorities to take positive steps, but it is more concerned with political and civil rights than with economic and social rights. Notably, it has been held that Article 8, which protects the right to a private and family life, "imposes no positive obligations to provide treatment".[31]

[25] See sections 1 and 3. Under the new Health and Social Care Act 2012, the Secretary of State will retain legal responsibility for duties under section 1, but a new public body, the NHS Commissioning Board, will have broad overarching duties to promote the comprehensive health service (other than in relation to public health) and clinical commissioning groups will take over responsibility for the duties under section 3. See the Health and Social Care Act (www.legislation.gov.uk/ukpga/2012/7/contents/enacted/data.htm).

[26] Although, under the NHS Constitution 2010, patients "have the right to drugs and treatment that have been recommended by NICE [the National Institute for Health and Clinical Excellence] for use in the NHS, if your doctor says they are clinically appropriate for you" (NHS Constitution, p. 9). (see www.nhs.uk/choiceintheNHS/Rightsandpledges/NHSConstitution/Documents/nhs-constitution-interactive-version-march-2010.pdf).

[27] See *R* v. *Secretary for State for Social Services ex parte Hincks* (1980) 1 BMLR 93. It was held in this case that the previous NHS Act 1977 did not create "private" rights, but the Court of Appeal did not rule out the possibility of an action for judicial review of an alleged failure to comply with the relevant statutory duties if the Minister had acted either in bad faith, or overtly contravened the basic policy enshrined in the Act.

[28] *Associated Provincial Picture Houses Ltd.* v. *Wednesbury Corporation* [1948] 1 KB 223.

[29] See, for example, *R (Murphy)* v. *Salford Primary Care Trust* [2008] EWHC 1908; *R (on the application of Rogers)* v. *Swindon NHS Primary Care Trust* [2006] EWCA Civ 392.

[30] *R* v. *North Derbyshire Health Authority ex parte Fisher* [1997] 8 Med LR 327.

[31] *R* v. *North West Lancashire Health Authority ex parte A & B* [2000] 1 WLR 977, at 995, *per* Auld LJ. And, more recently, see *R (on the application of Condliff)* v. *North Staffordshire Primary Care Trust* [2011] EWHC 872, [2011] EWCA Civ 910.

It is therefore extremely unlikely that any individual with BIID would be successful in either claiming compensation for the National Health Service refusing treatment or legally compelling the provision of treatment.

However, there is also recourse to private law, which governs the relationship between the doctor and his patient. The doctor has a legal obligation to patients for whom he has assumed care (or where in a hospital, the hospital has assumed care), but is not under a legal obligation to assume that care.[32] He has a duty to take positive steps where these are appropriate, and the hospital trust for which he works will be vicariously liable for his negligent acts of either commission or omission[33] that cause the patient to suffer harm. In ascertaining whether a breach of duty has occurred, the law of negligence tests professional performance by considering whether the doctor has behaved in accordance with reasonable medical practice. This principle is based on the *Bolam*[34] test, according to which:

> A doctor is not negligent if he adopts an approach acceptable to a respectable body of medical opinion (even if it is not in accordance with majority medical opinion) and if he exhibits the standard of care and skill to be expected of the averagely competent medical practitioner (even if it falls short of the standards of the leaders of the profession).[35]

Although the principle of respect for autonomy may give rise to the right of the patient to refuse treatment, English law is very limited in its ability to support the patient who demands treatment.[36] The law does not compel doctors to act in a manner contrary to their personal conscience,[37] and as many doctors do not accept that amputation is appropriate for BIID, it seems unlikely that any individual could successfully demand surgery – it would be very difficult to establish that a

[32] *R* v. *Central Birmingham HA ex parte Walker* (1987) 3 BMLR 32; *R* v. *Central Birmingham HA ex parte Collier, The Times*, 6 January 1988.

[33] *Bolitho* v. *City and Hackney HA* [1998] AC 232 is an example of a case where a failure to act was held to be negligent, although causation was not proven.

[34] *Bolam* v. *Friern Hospital Management Committee* [1957] 1 WLR 582.

[35] As presented in J. Munby, 'A Duty to Treat? A Legal Analysis', in S. Smith and R. Deazley (eds.), *The Legal, Medical and Cultural Regulation of the Body* (Aldershot: Ashgate, 2009), p. 179. Note that the "*Bolam* test" has been modified by the judgment in *Bolitho* (n. 33, above).

[36] See *R (on the application of Burke)* v. *General Medical Council* [2005] EWCA (Civ) 1003 (at [31]): "Autonomy and the right of self-determination do not entitle the patient to insist on receiving a particular medical treatment, regardless of the nature of the treatment. In so far as a doctor has a legal obligation to provide treatment, this cannot be founded simply upon the fact that the patient demands it."

[37] Note, for instance, the doctor's right to conscientious objection under section 4 of the Abortion Act 1967.

refusal to amputate constitutes a breach of duty according to professional standards. Moreover, proving causation would also be problematic, as the likely conclusion would be that it is the BIID condition that is causing the patient to suffer harm, rather than the doctor's refusal to amputate.

Another question is whether a patient can demand an alternative treatment that the doctor does not offer. For example, if a doctor offers an individual with BIID the choice of cognitive behavioural therapy or pharmacological treatment, could he or she decline both treatments but demand amputation, a procedure that the doctor has not offered? It is unlikely that such a demand could be legally enforced.[38]

Legal action against the surgeon for performing an amputation

This is the situation where there is probably the greatest uncertainty in the ramifications of surgery for BIID.[39] In theory, legal action could be taken against a surgeon amputating a healthy limb, for he would *prima facie* commit an offence contrary to section 18 of the Offences Against the Person Act 1861 (OAPA), which provides:

> Whosoever shall unlawfully and maliciously by any means whatsoever wound or cause any grievous bodily harm to any person ... with intent ... to do some ... grievous bodily harm to any person, ... shall be guilty of an offence, and being convicted thereof shall be liable ... to imprisonment for life.

A surgeon amputating a limb may be said to cause grievous bodily harm (GBH) or 'really serious bodily harm' to his patient.

At present, there is no precedent of any attempted prosecution of a surgeon in the United Kingdom for such an amputation. However, there remains the possibility that the Crown Prosecution Service (or Procurator Fiscal in Scotland) could institute a prosecution on the grounds of causing grievous bodily harm with intent under the OAPA offence referred to above, even if there was no complaint from the patient. There is precedent in this regard in the case of *R* v. *Brown*, where individuals were injured in a cooperative *and consensual* sado-masochistic exercise.[40] The Crown Prosecution Service instituted a successful prosecution, despite there being no complaint from the injured parties.[41] A surgeon who carries out an amputation on a BIID

[38] See n. 36, above.

[39] J. Johnston and C. Elliott, 'Healthy Limb Amputation: Ethical and Legal Aspects' (2002) 2 *Clinical Medicine* 431.

[40] *R* v. *Brown* [1994] 1 AC 212.

[41] R. Green, 'Serious Sado-Masochism. A Protected Right of Privacy?' (2001) 30 *Archives of Sexual Behaviour* 543.

patient could be regarded as causing grievous bodily harm, as an amputation produces a major and irreversible change to the patient. The procedure, however, is carried out with full informed consent and at the request of the patient. In contrast to the sadomasochistic exercise described above, there is a known, predictable and planned outcome, which has none of the uncertainty of spontaneous injury created in sexually aroused situations. Moreover, the law recognises that there are circumstances in which the public interest justifies contact causing or risking bodily harm. Contact sports are one example. Proper medical treatment, including reasonable surgery, is another, and in *R* v. *Brown*, Lord Mustill made it clear that if the law did not recognise that surgery was a good reason for bodily harm, many surgeons would be criminals. He argued that proper medical treatment with consent was lawful and in a category of its own.[42] The basis for any prosecution for amputation in cases of BIID would have to rest on advancing expert evidence that such surgery was improper.

Globally, I am aware of only one prosecution in a case of amputation for BIID and that was the case of John Ronald Brown, who carried out an above-knee amputation on Philip Bondy, a BIID sufferer, in a clinic in Tijuana, Mexico. A few days later, Bondy was found dead of gas gangrene in a hotel room in California. Brown was eventually convicted of second degree murder and sentenced to fifteen years to life in prison.[43] Brown was an unlicensed, seventy-six-year-old practitioner, who had already been imprisoned for carrying out surgery without a licence. In this case, the conviction was related to negligence, incompetence and unlicensed practice, rather than the actual indication for the amputation. I believe that *properly* conducted amputation for BIID, in the case of an appropriately assessed, prepared and consenting competent patient, should not be regarded as an illegal act.

Whilst a patient is extremely unlikely to complain in the short term, in the long term if they are unhappy with the outcome (particularly if age and increasing infirmity make life more difficult), they may pursue the surgeon for other forms of legal redress. For example, they might have a claim under the tort of negligence for a failure to obtain informed consent if certain risks are not disclosed,[44] or argue that the surgery was unlawful and therefore their consent was of no effect.

[42] And see Law Commission, Consultation Paper No. 139, *Consent in the Criminal Law* (London: HMSO, 1995) and *Airedale NHS Trust* v. *Bland* [1993] AC 789 at 891.
[43] See *People* v. *Brown* (2001) 91 Cal. App. 4th 256.
[44] See *Sidaway* v. *Bethlem Royal Hospital Governors* [1985] AC 871; *Pearce* v. *United Bristol Healthcare Trust* (1999) 48 BMLR; *Chester* v. *Afshar* [2004] UKHL 41.

Surprisingly, there are no specific guidelines produced by the GMC with regard to the role of surgery or providing guidance for those working with such patients. When I first planned an elective amputation for a BIID sufferer, I approached the GMC for advice, and the only response that I received was that I should consult their document, *The Duties of a Doctor*,[45] and proceed on the basis of the information contained in that document. They also commented that if there were a formal complaint with regard to the procedure, they would launch an investigation.

There is also no research base for the provision of surgery. As Carl Elliot points out,[46] there are no licensing procedures for the introduction of new surgical procedures or, for that matter, new psychotherapeutic procedures, in contrast to the stringent licensing requirements for new drugs and devices. As surgeons, we are having our work increasingly scrutinised by our peers, our employers and our patients, as well as focus groups, and it is therefore important that any new form of non-pharmacological interventional therapy should be stringently assessed before introduction. In most cases, any new form of therapy is only used under carefully audited conditions in some form of clinical trial. In my view, some form of controlled trial of surgery for this group of individuals would be extremely difficult to carry out, as, to give meaningful results, there would need to be a large population treated and a follow-up over many years. Recruitment of a sufficient number of patients is a problem that besets all trials of therapy for rare conditions.

The introduction of a formal licensing procedure for new interventional therapies would be ideal, but this would inevitably lead to delays in introduction, and with the rate at which surgery is developing at present it would be a logistical nightmare. Nevertheless, I would argue that there should be some form of recognised official assessment of new procedures before widespread introduction, although not to the extent of the current licensing of drugs, which is a very prolonged, complex and expensive procedure.

The future

The future treatment of patients with BIID is dependent on rational debate as to what is best for the patient and should not be confined by prejudice and rigid views. Some form of long-term follow-up study of

[45] General Medical Council, *Duties of a Doctor: The Duties of a Doctor Registered with the General Medical Council* (see www.gmc-uk.org/guidance/ethical_guidance/7162.asp).
[46] C. Elliot, 'A New Way to Be Mad' (2000) 283(6) *Atlantic Monthly* 72.

those who have achieved amputation would help, but it is unlikely that any form of controlled trial of therapy would be achievable. I suspect that no BIID sufferer would be happy to be in the "control" (non-operated) group.

I believe that we can learn a great deal from the development of treatment guidelines for patients with Gender Identity Disorder (GID). Patients with GID faced exactly the same situation in the 1970s, when there was extensive debate about the rationale for surgical treatment. The similarities with BIID are compelling[47] and there is growing evidence that both are due to a neuropsychological anomaly. With time, it has become accepted that surgery is appropriate for GID and the Harry Benjamin Guidelines for treatment[48] have been developed. There is a desperate need to provide some form of therapeutic guidelines, along the lines of the Harry Benjamin Guidelines, for those with BIID.

The Harry Benjamin Guidelines specify very precisely the procedures for assessment and diagnosis of patients with GID and the professional standards of those involved. They specify the use of psychotherapeutic interventions, hormonal therapy and specify a period of real-life experience before surgery is offered. Living convincingly as an amputee would be difficult, unlike living as a member of the opposite sex, so this criterion could prove to be a stumbling block. However, I do know of one man who wishes to have both arms amputated and who has lived as an armless man, with all the necessary supporting services, for over ten years. This would more than fulfil criteria such as those in the Harry Benjamin Guidelines.

In my view, the law has little role in the development of a guideline. A guideline needs to be developed by a multidisciplinary team of concerned and experienced individuals. It must take account of all available evidence and work within current legal constraints. In the absence of evidence, a consensus view needs to be developed. What the law should do, however, is respond to and support the implementation of the guideline.

Conclusions

As indicated in the title of this essay, the debate rages around perceptions of the individual with BIID held by the general public and the health professions. The general public and many health professionals regard

47 Lawrence, 'Clinical and Theoretical Parallels'.

48 *The Harry Benjamin International Gender Dysphoria Association's Standards of Care for Gender Identity Disorders* (6th version) (2001) (see http://wpath.org/Documents2/socv6.pdf).

amputation as a devastating and disabling condition, but in certain conditions, amputation can be life changing. Patients who have intractable pain due to ischaemia (a lack of blood supply) gain enormous relief from their symptoms and often find their life is transformed. It is difficult to comprehend that amputation can improve one's psychological wellbeing, but for the BIID patient this positive effect would be just as real as for the patient with a dead limb.[49]

The wish to be an amputee is irrational and completely contrary to normal human desires, so BIID patients are regarded as severely psychiatrically disturbed. The perception is that as these patients are so mentally ill, they have to be protected from the folly of achieving amputation. Indeed, some patients have even been detained under compulsory orders. However, careful psychiatric assessment has failed to identify any significant mental illness. The sufferers themselves recognise that their desire is "nuts", but they do not accept that this is a mental illness. My experience of these patients is that they are well-educated, intelligent people, who are as bewildered by their feelings as the rest of the population. But their feelings are very real and personal. They fully understand the consequences of what they are requesting, have in-depth knowledge of amputee rehabilitation, and understand the lifestyle of amputees. To them, amputation is the only source of relief. They are fully competent to make informed decisions about their treatment and there is no concern about competency to consent. If this is a neuropsychological condition, the various forms of therapy available for psychiatric illness are unlikely to be effective. The possibility of physical therapy is currently being explored, but seems to provide, at best, only very short-term relief. Anecdotally, surgery produces complete relief of symptoms without creating dependency.

When we consider the ethical issues that this condition raises, it seems clear to me that if we feel that respecting individual autonomy and allowing competent individuals control over their lives is generally the right thing to do, then we should allow this surgery. Those with BIID appear to have the requisite mental capacity to consent to this procedure and they do well after the procedure, to the point where I argue that the procedure is likely to be resource-neutral, as they cease to use psychiatric resources, become more productive citizens and prefer to be recognised as amputees. But while amputation in such cases seems to be completely compatible with central ethical issues, such as respect for individual

[49] Although I recognise that in the latter case, the undesired death of the limb and its subsequent amputation are likely to have a negative impact on the individual's psychological wellbeing.

autonomy and concerns for justice, this is not usually the view taken by the population in general. Herein lies the problem of perception. Amputation is regarded as harmful to the individual's interests and incompatible with concerns for justice by the population at large. However, despite this contrary public perception, it seems to me that respect for individual autonomy of the patient takes precedence. Thus, from an ethical standpoint, I think that amputation should be offered.

The legal situation, however, is unresolved. The legal systems in the UK are based on precedent and, so far, there is no precedent for this particular situation. Prosecution is an expensive option and it is difficult to see how prosecution could help either the community or the individual concerned. The GMC, as I have shown, is not really interested in making a decision on this condition as it is relatively rare. The GMC's response tends to be reactive rather than proactive in such circumstances, as I found in my communications with them. Although there can be no compulsion to provide amputation, there should be an agreement that willing surgeons are allowed to provide amputation to suitable patients without fear of prosecution.

In my view, the only way forward for BIID sufferers is to develop some guidelines along the lines of the Harry Benjamin Guidelines for Gender Identity Disorder by a group of interested specialists and sufferers. This must be combined with the opportunity for surgical intervention as part of a research project. The development of a formal guideline document could act as a basis for therapeutic rationalisation and legal recognition of the condition and its therapy.

During the considerable publicity that followed the amputations I carried out in the late 1990s, I received a letter from a seventy-five-year-old lady who had had a leg amputation whilst in her twenties. She said, "Living with an amputation is nothing compared to living a lifetime of mental torment."

6 Risky sex and 'manly diversions': contours of consent in HIV transmission and rough horseplay cases

David Gurnham

Introduction

After the House of Lords' effective criminalisation of sadomasochism in *R* v. *Brown*,[1] there still remained questions about exactly how matters of sex, violence and risk intersected in setting the boundaries of lawful consent. In the jurisprudence on the reckless infliction of grievous bodily harm under the Offences Against the Person Act 1861 (OAPA), the courts seem to have eschewed a consistent approach in favour of different standards depending on the types of behaviour and context. 'Rough and undisciplined horseplay' is considered to be lawful, even when bodily harm is foreseen and consent wrongly presumed, provided that there is no intention to hurt. However, in the case of unsafe sexual activity where the defendant (D) knows he is infected with HIV and actually transmits it to the victim (V), D is guilty of recklessly inflicting grievous bodily harm unless V can be shown to have given 'informed' consent to that risk. The question that I want to address in this essay is whether this obvious discrepancy can be defended on principle, or if it is likely that the courts will soon follow their own rulings (i.e. from the HIV judgments) and take a harder line on consent in rough horseplay cases too. I suggest here that the courts' continued toleration of the discrepancy arises from the different sorts of gender assumptions that underlie the two species of case law. This essay will draw out the ways in which essentialist traditional assumptions about men and women have been relied upon in particular cases, which have in turn shaped legal notions of autonomy, consent and equality. I shall show how the selective engagement by modern courts with historical authorities has produced these two inconsistent readings of the offence of maliciously inflicting grievous bodily harm (GBH) under section 20 of the OAPA. On the one hand, the celebration of violent 'manly diversions' is embraced within modern

[1] *R* v. *Brown* [1993] 2 All ER 75.

notions of equality (producing no 'victims' of 'friendly' violence), while at the same time the historic legal inferiority of married women is repudiated in order to justify a much stricter treatment of consent in HIV cases, distancing modern law from the inequalities of the past. This is a fissure in the treatment of consent that has opened up because of an antagonism, not between autonomy and paternalism, but between two views of autonomy that have emerged from criminal legal history. It is not that there is inconsistency in the law on when consent is 'enough' to establish lawfulness, but rather that the way in which a person is *deemed to have consented* differs according to the type of harm involved and the context of its infliction.

Ascertaining a victim's consent: when is it presumptuous to presume?

Consent to grievous bodily harm inflicted through 'rough and undisciplined horseplay'

In *R* v. *Jones*[2] and *R* v. *Aitken*,[3] the defendants successfully appealed against their convictions for maliciously inflicting GBH under section 20 of the OAPA, on the basis that the activities in question were lawful, the harm accidental and the victims either consented to the risks or were believed by the defendants to have so consented.[4] Rough horseplay had been recognised as lawful in the nineteenth century,[5] but the Court of Appeal in both cases relied instead on the Lord Chief Justice's fortuitous use of the word 'etc.' in his list of lawful (though potentially harmful) activities in *Attorney-General Reference (No. 6 of 1980)*,[6] as including 'properly conducted games and sports, lawful chastisement or correction, reasonable surgical interference, dangerous exhibitions, etc.'.[7] The Court decided that so long as the violence could be regarded as falling within the bounds of mere horseplay and that bodily harm was not the motive, then the risks involved and the harm actually caused could be treated as lawfully consented to, either in fact or by presumption. In *Jones*, the defendants were six boys aged between fourteen and seventeen, who injured two other boys by 'grabbing', 'marching' and finally

2 *R* v. *Jones* (1986) 83 Cr App R 375. 3 *R* v. *Aitken* [1992] 1 WLR 1006.
4 According to section 20 of the OAPA, it is an offence punishable for up to five years to 'unlawfully and maliciously wound or inflict any grievous bodily harm upon any other person'. The Act is silent on the relevance of consent as a defence.
5 *R* v. *Bruce* (1847) 2 Cox, C C 262.
6 *Attorney-General Reference (No. 6 of 1980)* (1981) 73 Cr App R 63.
7 Ibid., at 66.

throwing them up high in the air and letting them fall to the ground, causing a ruptured spleen and a fractured arm.[8] All the defendants claimed that it was a joke and that the victims thought so too. Whether or not the victims really did share the joke did not seem to interest the Court, and although McCowan J in the Court of Appeal admitted that the victims were in fact not consenting, but instead 'protested and claimed that they were being hurt',[9] he took the defendants' assertion at face value. The parties in *R* v. *Aitken* were newly qualified RAF pilots, and were convicted of inflicting 35 per cent burns, amounting to grievous bodily harm, on a colleague by setting him alight while he slept, at the end of a night of heavy drinking. The question was the application of consent, and Cazalet J in the Court of Appeal states the facts in terms that at first do seem to suggest that there was no consent:

> [Gibson] was staggering about, was slurring his words, and said that he was going to bed. He left the bar. The three appellants followed him. ... They took hold of, and grabbed him. He resisted, but because of the amount which he had drunk he struggled only weakly. ... The fuel was ignited. ... Aitken that night said that Gibson had gone up 'like a torch'.[10]

As established in *Jones*, presumed consent is a rather different matter to actual consent. Although not permitted to take the defendants' intoxication into account when considering their recklessness, it is clear from Cazalet J's approval of the defence's construction of the facts that the context of the horseplay pointed towards acceptance of presumed consent:

> The fact that Gibson struggled, albeit weakly through drink, to avoid the attentions of the three during the incident in question should not, it was submitted on the appellants' behalf, be taken in isolation. The totality of the circumstances, his knowledge of the course which celebration evenings such as the one in question was likely to take and his continued presence with the others demonstrated an acceptance by him that horseplay of the nature perpetrated upon him might well take place.[11]

But as Cazalet J himself had already made clear, Gibson had left the bar and gone away to sleep. He was only a 'continued presence with the others' because they had followed him, intending to set him on fire. This approach to finding consent was approved by Lord Woolf CJ in *R* v. *Barnes*[12] in the context of a deliberate off-the-ball foul in an amateur

8 *Jones*, *per* McCowan J, 377.
9 All quotes from *Jones* (1986) 83 Cr App R 375, at 377.
10 *Aitken*, at 1009–1010. 11 Ibid., 1019–1020.
12 *R* v. *Barnes* [2004] EWCA 3246.

football match: could a foul of the kind that broke the victim's ankle be 'anticipated in a normal game of football', or was it 'something quite outside what could be expected'?[13] Although he does not cite *Jones* or *Aitken* as authorities, Lord Woolf CJ's approach clearly implies the same thing, namely that V's consent in taking part in the match at all has a certain scope, the bounds of which are matters of public policy, and the question for the jury is to consider whether injuries sustained may be deemed to fall within that scope.

Consent in cases of reckless transmission of HIV through sexual intercourse

Unlike the rough horseplay and games cases, the question of consent to the risk of HIV infection following unprotected sexual intercourse has received very careful judicial attention. The leading cases of *R* v. *Dica*[14] and *R* v. *Konzani*[15] carve out HIV infection as a kind of harm requiring specific consent, and thus quite different from other forms of GBH inflicted unintentionally through risky behaviour. In both cases, an HIV-positive man was convicted under the same offence as prosecuted in the 'horseplay' cases described above, for recklessly infecting a number of women. Dica's conviction was quashed on the basis that the trial judge had wrongly instructed the jury that they should convict irrespective of whether he disclosed his HIV infection, on the grounds that, following *Brown*, GBH inflicted for sexual gratification cannot be made lawful by V's consent.[16] Judge LJ in the Court of Appeal distinguished the instant facts from *Brown* insofar as it lacked the violence and 'cruel' motive of sadomasochism and thus the mere risk that serious harm might occur could not preclude consent from ever being acceptable as a defence. *Dica* does give some scope for presumed consent, as Judge LJ anticipated that people who consented to unprotected 'casual sex between complete strangers' should be deemed to have also consented to the risk of HIV.[17] In such cases, Judge LJ pragmatically accepted that criminal law would be ineffective at affecting behaviour, and if the victim could have (but failed to) protect themselves from the risk of infection, they were on sufficiently equal terms with the defendant as to be considered a victim of their own lifestyle rather than of a criminal offence. In the later judgment of *Konzani*, Judge LJ significantly narrowed the defence: unless victim was in a position to give 'informed' consent – that

[13] Ibid., para. 28. [14] *R* v. *Dica* [2004] 2 Cr App R 467.
[15] *R* v. *Konzani* [2005] 2 Cr App R 14. [16] *Dica*, *per* Judge LJ, 480, para. 46.
[17] Ibid., 480, para. 47.

is, consent to unprotected sex *knowing* that the defendant was HIV positive – there was no consent to the risk of HIV, irrespective of the nature of the sexual relationship.[18] In the absence of such a concrete reason to believe it,[19] defendants are not entitled to presume that anyone who has unprotected intercourse implicitly consents to all the attendant risks, meaning that the scope of the victim's consent is now deemed to extend only to the sex and not to its possible adverse effects.

Adopting the terms used by Joel Feinberg, the *Konzani* reconfiguration of consent means that HIV-transmission victims may now be treated as though they were not merely 'harmed', but also 'wronged' by the defendant who allowed them to put themselves into a vulnerable position in ignorance of a potentially devastating risk that negates their opportunity to make a fully autonomous choice.[20] Using Feinberg to describe what he regards as the rightful domain of the criminal law, Charles Erin argues that no matter how severely harmed a person is, they are not *wronged* (and thus should not be regarded as the victim of a criminal offence) unless what is done to them exceeds the scope of their consent.[21] But how widely ought we to draw this scope? On the one hand, to treat an adult with knowledge of the general risks of unprotected sexual intercourse as fully autonomous would mean we would expect them to take full responsibility for what Erin calls 'stupid choices'.[22] Thus Weait argues that couples who 'pursue sexual pleasure outside established and socially legitimized relationship structures may be assumed to have consented to the risk of infection if it occurs'.[23] Similarly, Bennett *et al.* have argued that to withhold information about one's HIV-positive status from a sexual partner is both harmful *and* wrongful, but only in cases where one understands the other to have made an 'assumption of negativity'.[24] Like the earlier

[18] *Konzani*, *per* Judge LJ, para. 41. [19] Ibid., para. 44.

[20] For example, D. Omerod and M. Gunn, 'Criminal Liability for Transmission of HIV' (1996) 1 *Web Journal of Current Legal Issues* 23; J. Herring, 'Mistaken Sex' [2005] *Criminal Law Review*, 511.

[21] C. A. Erin, 'The Rightful Domain of the Criminal Law', in C. A. Erin and S. Ost, *The Criminal Justice System and Health Care* (Oxford University Press, 2007) p. 237, at p. 246, quotes with approval Feinberg on this: 'The harm principle will not justify the prohibition of consensual activities even where they are likely to harm the interests of the consenting parties; its aim is to prevent only those harms that are wrongs.' (Quoted passage from J. Feinberg, *Harm to Others* (New York: Oxford University Press, 1984), pp. 35–6.

[22] Ibid.

[23] M. Weait, 'Harm, Consent and the Limits of Privacy' (2005) 13 *Feminist Legal Studies*, 97, 117.

[24] R. Bennett, H. Draper and L. Frith, 'Ignorance Is Bliss: HIV and Moral Duties and Legal Duties to Warn' (2000) 26 *Journal of Medical Ethics* 9, 12–13.

judgment of Judge LJ in *Dica*, Bennett *et al.* are prepared to distinguish between 'casual' and 'responsible' relationships, arguing that non-infected parties would only be entitled to make such assumptions once a 'trust relationship' had developed, and that where the victim had made a 'rash' assumption of negativity, the defendant would not be considered to have wronged them.[25] On the other hand, we might say, as the *Konzani* Court did, that in every act of sexual intercourse, a person can only be taken to have consented to that which s/he was made aware of at the time.[26] However, such an approach requires us to think carefully about whether there is something particular about this kind of risk-taking that warrants such special treatment by the criminal law.

Three types of risky behaviour, three legal regimes: horseplay, HIV and 'vigorous' sex

In contrast to the duty on D to disclose his HIV infection before he can rely on V's consent as a defence, a defendant who injures another during a bout of 'rough horseplay' need only have had 'honest' belief in the victim's consent. Although horseplay may expose victims to grave dangers, judicial rhetoric has so far shied away from attaching the same aggressor–victim rhetoric to its treatment of consent as compared to cases involving sexual activity. The consent regime applying to horseplay derives from its historic recognition as lawful activity (and hence an exception to the general prohibition of dangerous behaviour) in *Donovan*,[27] *Attorney-General Reference* and *Brown*. Horseplay has always enjoyed a legitimacy denied to sexually motivated behaviour, on grounds of public interest. *Dica* and *Konzani* agree that although risky activity is not necessarily unlawful simply because of sexual motivation, sexual expression or gratification are not in themselves considered good enough reasons to allow a victim's consent to justify the reckless infliction of harm.[28] Other cases have distinguished sexual and non-sexual activity to make assessments of how likely it was that the victims actually consented

[25] Ibid., 13. It is worth noting, however, that some critics have claimed that such a view is still recognised under *Konzani*. For instance, Vanessa Munro suggests that the Court of Appeal in *Konzani* may have been influenced by the fact that 'the sexual intercourse in question took place in the context of more than casual relationships and involved women who might be thought liable to be particularly trusting' (V. Munro, 'On Responsible Relationships and Irresponsible Sex: Criminalising the Reckless Transmission of HIV *R* v. *Dica* and *R* v. *Konzani*' (2007) 19 *Child and Family Law Quarterly*, 112). See also S. D. Pattinson, 'Consent and Informational Responsibility' (2009) 35 *Journal of Medical Ethics*, 176, 178.

[26] Erin, 'The Rightful Domain', p. 256.

[27] *Donovan* (1934) 25 Cr App R 1; [1934] KB 498.

[28] See *Dica* 480, para. 46.

to a dangerous activity. For example, in *Boyea*,[29] involving a male defendant and a female victim (both of whom were intoxicated) in a sexual encounter that resulted in injuries to the woman's vagina, the defendant was denied the defence of consent on appeal because 'it is in our view *inconceivable* that Miss Collins would have consented to the injuries which were in fact inflicted on her'.[30] Although the question of whether Miss Collins consented was disputed between the parties, this was not referred back to a jury for consideration. Rather a lack of consent was determined by the Court of Appeal itself, and this finding was emphasised in Glidewell LJ's reconstruction of the facts as attributed to the complainant, vividly recreating in text the violence of the sexual act: He '*pushed* her onto the bed, ... *ripped* off her underclothes, ... put his hand *round her throat* ...',[31] and so on. To some readers it might be inconceivable that anyone could be said to have been consenting when set alight in their sleep or thrown ten feet in the air and dropped on the ground. As in the horseplay cases, the *Boyea* Court implies that there is a scope for making presumptions about what V's consent actually entailed, but decides that on the facts, D's actions here are simply beyond the pale.

Complicating this picture further, comparing the *Konzani* position on HIV transmission to that governing both sexual and non-sexual risky behaviour reveals that there are three quite different legal orders operating under section 20. In *R* v. *Meachen*,[32] the defendant's claim that V (a woman) had consented to taking the 'date-rape' drug gamma-hydroxybutyrate (GHB) and then to vigorous sexual activity was sufficient for the Court of Appeal to quash his conviction under section 20. This was despite her suffering 'acute splitting of the anal canal area extending to the rectum ... so severe that a colostomy was performed',[33] because the injury itself may not have been foreseen or intended by him. If D had admitted that he had foreseen the danger of injury to the woman, then his conviction under section 20 would have been upheld because the activity itself was sexually motivated, and, as noted earlier, unlike horseplay, sexual gratification has never been deemed to be a 'good reason' to expose another to a foreseeable risk of injury, irrespective of the victim's consent. Following *Konzani*, however, had *Meachen* been HIV positive and the harm the victim suffered been HIV infection, the lack of extreme violence would mean that the defence of consent would apply even if bodily harm was anticipated, but only fully informed consent to the

29 *R* v. *Boyea* (1992) 156 JP 505.
30 Ibid., 513 (emphasis added).
31 Ibid., 506 (emphasis added).
32 *R* v. *Meachen* [2006] EWCA Crim 2414.
33 Quoted facts, para. 11.

specific risk of HIV transmission would suffice, not merely consent to unprotected sex. In my view, it is wrong for law to categorise these three scenarios separately, either on the basis of whether the harm suffered was serious infection or serious injury, or whether the motivation was sexual or non-sexual. All three arise from *inherently* risky activity; all involve potentially foreseeable harm; and in all three instances the harm may be inflicted 'accidentally' in a lay sense. The courts insist the 'public interest' is a reliable yardstick of the legal rule to be applied, but without further justification of judicial understanding of public interest, it remains unclear why the meaning of 'accidental' should depend so much on the motivation for the dangerous activity, or on whether the harm takes the form of infection or injury. This distinction alone is difficult enough to comprehend from a liberal perspective, but the clumsiness of the public interest distinction in the context of risky sexual activity is made even clearer in the HIV cases. Judge LJ in *Dica* specifically noted that 'the problems of *criminalising* the consensual taking of risks like these include the sheer impracticability of enforcement and the haphazard nature of its impact'.[34] He judged that to apply the same rules to HIV transmission as had been applied to cases involving sexually inflicted bodily injury (such as *Boyea*, *Brown* and *Meachen*) raises enforcement difficulties and would extend to convicting those he considered responsible citizens acting for 'good reasons', such as 'the mother or father of a child suffering a serious contagious illness, who holds the child's hand, and comforts or kisses him or her goodnight'.[35] Judge LJ also expressed his discomfort at an application of legal principle that would criminalise the Catholic couple whose moral objections to contraception would require them knowingly to run the risk that the infected partner might infect the healthy one, or the similarly informed and infected couple desperate for a naturally conceived child.[36] The effect of the *Konzani* duty to disclose is therefore that, while V who consents to unprotected sex is now recognised as a victim of crime if the HIV is transmitted, even if they are fully aware of the risks, X, who runs the risk with *specific* knowledge of the risks, is not.

Reserving a requirement of 'reasonable belief' for HIV cases arguably brings this seam of section 20 offences into line with the Sexual Offences Act 2003, which requires juries in rape trials to consider whether a defendant's belief in the complainant's disputed consent was reasonable. This has caused some critics to consider whether the law has now begun to confuse the elements of sexual and non-sexual offences,

[34] *Dica*, para. 51 (emphasis added). [35] Ibid., para. 51. [36] Ibid., para. 49.

despite the courts making it clear that being deceived as to the 'nature and quality of the act' (rape) does not include deception as to D's HIV status.[37] However, the important point for present purposes is the particular issues of public policy that the HIV judgments bring to bear. The 'special' legal status of consent as regards HIV transmission arises because the activity through which the infection is transmitted – consensual sexual intercourse or other sexual activity – is not itself against public policy, but when combined with violence or catastrophic infection may carry consequences serious enough to interest the criminal law. We have seen that, on grounds of public policy, it is not possible to give lawful consent to the risk of injury merely for sexual gratification, but since there are (in Judge LJ's view) important non-sexual reasons to consent to the risk of HIV, Judge LJ's innovative 'third way' allows such risks to be run without compromising law's commitment to public policy or resorting to presumed consent. In the next section, I offer a historical-rhetorical explanation as to why the law on presumed consent has diverged in this way.

Consent, gender and precedent: a historical view

Could these inconsistencies in the criminal law's approach to risky behaviour be explained by differences in the way social values have intersected with the developing law? The groundwork for the more radical judgment of *Konzani* was undertaken by Judge LJ in the earlier case of *Dica*, in his pointed and much-laboured refusal to follow the nineteenth-century authority of *R* v. *Clarence*.[38] In this now much-derided judgment, it was held that a man who knowingly exposed his (unknowing) wife to gonorrhoea through consensual sex could not be guilty of an offence because there was no such thing as an indecent assault committed against a married woman by her husband, and without such an assault there could be no offence of maliciously inflicting bodily harm.[39] In *Dica*, *Clarence* plays an important role in Judge LJ's construction of a new consent regime for HIV cases. However, I would

[37] *R* v. *B* [2006] 1 WLR 1567 at 1571, para. 17. See Law Commission, *Consent to Sex Offences: A Report to the Home Office Sex Offences Review* (London: TSO, 2000), p. 47, para. 5.21. But see D. Warburton, 'Rape: Consent; HIV' (2007) 71 *Journal of Criminal Law*, 116; Munro, 'On Responsible Relationships'.

[38] *R* v. *Clarence* (1888) 22 QBD 23.

[39] Ibid., *per* Pollock B, 63–64: 'The husband's connection with his wife is not only lawful, but it is in accordance with the ordinary condition of married life. . . . The wife as to the connection itself is in a different position from any other woman, for she has no right or power to refuse her consent.'

question Judge LJ's own asserted view that this importance derives from it being 'indistinguishable' from *Dica*.[40] I suggest instead that *Clarence* is used simply to show what modern law has left behind – that is, in repudiating it, Judge LJ was able to use his own judgment to position modern law as being on the side of equality and autonomy. Judge LJ makes no mention of *Jones* or *Aitken* in his judgment in *Dica*, despite those decisions having established the surely relevant point that consent to a risky but lawful activity may provide a defence to a charge of recklessly inflicting bodily harm when the risk inherent in that activity materialises. Instead, he devotes fifteen paragraphs to repeating the criticisms made about *R* v. *Clarence* in twentieth-century cases on the 'questions' of the status of married women and whether or not a violent assault is necessary to find there to have been an offence committed. Judge LJ condemns the 'irrevocable privilege' of a husband over his wife's body and 'the artificial notion that sexual intercourse forced on an unwilling wife by her husband was nevertheless bound in law to be treated as if it were consensual intercourse'.[41] But *Clarence* had already been decisively overruled on this point by the landmark House of Lords marital rape judgment in *R* v. *R*.[42] Having established that English law had indeed already rejected the notion that there was no such offence as sexual assault by a husband against his wife, Judge LJ moves on to make a similar criticism of the *Clarence* view that in the absence of assault there can be no conviction under section 20 of the OAPA.[43] Given legal developments since 1888, therefore, what *Clarence* had to say about the legal impossibility of a married woman being indecently assaulted by her husband, and whether a person could be charged under section 20 without an assault, are clearly so outmoded as to raise certain questions about the attention that Judge LJ's judgment gives them. Judge LJ defends his decision to focus so much attention on *Clarence* on legal grounds: 'If Clarence remains authoritative, this case is indistinguishable and therefore this appellant should not have been convicted. His convictions, like Clarence's, would have to be quashed.'[44] However, since *Clarence* was already obsolete on the relevant legal questions, the more compelling case for its inclusion is surely to be made, not on legal grounds, but rather in terms of the social and cultural assumptions about sexual relationships that underlie it.

40 *Dica*, *per* Judge LJ at 471, para. 17.
41 Ibid., at 472, para. 19.
42 *R* v. *R* (1992) 94 Cr App R 216.
43 Namely, in three important cases: *R* v. *Wilson (Clarence)* (1983) 77 Cr App R 319 *per* Lord Roskill 329 and 360; *R* v. *Chan-Fook* (1994) 99 Cr App R 147; *R* v. *Ireland; R* v. *Burstow* [1998] 1 Cr App R 177.
44 *Dica*, *per* Judge LJ, 471, para. 17.

Dica itself has nothing to do with either the status of married women or the scope of consent given by marriage, but in grounding his judgment in the outmoded nature of the older, apparently 'indistinguishable' case, Judge LJ underlines a break in legal history between social inequalities of the past and the equality and autonomy accorded to individuals in the present. His focus on the social and legal assumptions underpinning *Clarence* seems to be aimed at demonstrating that there is something particularly sensitive about a sexual relationship in modern social contexts that marks it out from other potentially 'risky' activities, and thus that there is a need for a particularly fine treatment of presumed consent in this context. In 1888, the necessary consent for a husband to exercise his 'conjugal rights' over his wife's body forever was provided when she took her vows of marriage; after that, consent seems to have been assumed for all subsequent sexual acts and their consequences. Judge LJ never addresses the horseplay cases in *Dica*, but a year later, counsel for Feston Konzani attempted to use *Jones* and *Aitken* to argue that consent to HIV infection was implicitly given when the victims consented to the inherently risky activity of unprotected sex with a man without taking proper precautions.[45] But Judge LJ's analysis in *Dica* had by then already established that the 'right' of a defendant to rely on fictive or constructed (rather than actual) consent in this particular context was now an unjustifiable throwback to an age of conjugal entitlement. Thus, in *Konzani*, Judge LJ needed only to refer the defence counsel to the decision in *Dica* in rejecting the submission. The implication that I draw from *Dica* and *Konzani* together on this point is that the infection of another with HIV by sexual intercourse, without gaining 'informed' consent, raises questions about gender and justice that injury by horseplay does not. It is possible that Judge LJ saw in *Dica* an opportunity for an '*R* v. *R* moment' – that is, an opportunity to signal a clear distinction between the enlightened modern law (with its commitment to autonomy, equality, etc.) and the dark ages of legal inferiority of women in the nineteenth century.[46] However, if the central issue in determining criminal liability for sexual HIV transmission really were the scope of consent to sex, then ought the behaviour of *Dica* not to be characterised as rape, and charged, not under the OAPA, but rather under section 1 of the Sexual Offences Act 2003?[47] This is not the

45 *Konzani*, *per* Judge LJ, para. 36.

46 See D. Monk, 'Reckless Trials? The Criminalization of the Sexual Transmission of HIV' (2009) 156(July/August) *Radical Philosophy* (see www.radicalphilosophy.com/commentary/reckless-trials).

47 S. Pattinson, 'Consent and Informational Responsibility', 178. Assault by penetration, section 2 Sexual Offences Act 2003, also requires 'reasonable belief' in consent and is not explicitly gendered as rape is.

conclusion arrived at by Judge LJ, but his deployment of equality rhetoric through the denunciation of *Clarence* seems to imply a gender dynamic more commonly attributed to rape: that a man's liability under section 20 for inflicting HIV must depend on constructing legally effective consent in such a way as to recognise a woman's equal status and sexual autonomy.

Judge LJ's judgments in the HIV cases repudiate a particular moment of English legal history in order to emphasise the legitimising values underpinning modern law. Comparing these judgments to the horseplay and other cases involving dangerous non-sexual activity between men, we find a very different sort of method of reading precedent and drawing lessons therefrom. The judgments of *Jones*, *Aitken* and *Barnes* are almost entirely devoid of the victim rhetoric of *Dica* and *Konzani*. Indeed, the judicial rhetoric of these cases forcefully denies that there was ever a victim as such: the activity in question being not only lawful but positively healthy, either the complainant consented or else the defendant should be entitled to presume consent was given. Like *Dica*, the horseplay cases derive their view of what is required in the interpretation and application of statute through a response to ancient precedent. However, whereas Judge LJ deployed *Clarence* in order to signify the distance modern law has travelled towards gender equality, the horseplay cases resurrect eighteenth-century authority to establish continuity in the cultural assumptions underpinning the law. All the modern cases on the unintentional infliction of serious bodily harm recognise the early twentieth-century authority of *Donovan*. In that case, Swift J relied on Sir Michael Foster's 1792 treatise confirming the lawfulness of 'manly diversions, ... intended to give strength, skill and activity, [that] may fit people for defence, public as well as personal, in time of need'.[48] According to Foster, such manly diversions could include 'cudgels, foils and wrestling', which, if undertaken in the spirit of 'perfect friendship' would be lawful so long as all parties consented. Consenting to the risk of bodily harm was therefore a man's prerogative, and not something for the criminal courts to be involved with. Such militaristic eighteenth-century expectations that men will (and arguably by implication ought to) engage in such 'friendly' violence was used without gloss by Swift J and has been reproduced in appeal judgments ever since.[49] The Court of Appeal has recently repeated this view that manly sports require

[48] Ibid., *per* Swift J, 11/508, as quoted by McCowan J in *Jones*, at 378, citing Sir Michael Foster, *Crown Law* (3rd edn) (London: E. & R. Brooke, 1792), pp. 259–60 (see http://books.google.co.uk/books?printsec=frontcover&pg=PR1&id=4ZI0AAAAIAAJ#v=onepage&q=&f=false).

[49] For example, in *Brown*, *per* Lord Jauncey, 88.

leniency as to levels of violence. In the recent case of *Barnes*, Lord Woolf CJ warned that 'in highly competitive sports, conduct outside the rules can be expected to occur in the heat of the moment'.[50] It is a remark that echoes not only Sir Michael Foster's approval of 'manly diversions', but also Bramwell LJ's reluctance in a nineteenth-century case to 'decry manly sports'.[51] According to this traditional view of what it is to be and become a man, it is only natural for teenage boys (*Jones*) or newly qualified airmen (*Aitken*) to expect to put up with a bit of rough and tumble because it would equip them with the essential attributes of manhood. Like all lawful risky activities, the acceptability of rough games and horseplay surely now applies to men and women equally or not at all. However, it remains an idealisation of 'manhood', explicitly gendered, that continues to be used as the general signifier of equality between parties to 'friendly' violence.

Whether eighteenth- or nineteenth-century sources continue to be suitable to inform the social and normative content of criminal law is not at all a question about age. Rather, it is a question about how convincingly law deploys its rhetorical constructions in situating itself with regard to the relationship between the autonomy of individuals on the one hand, and the experience of the social world with all its dangers on the other. It is of course open to a future court to re-evaluate the application of consent in horseplay cases, as Judge LJ has done in HIV cases, namely to declare that the traditional celebration of dangerous 'manly diversions' is incompatible with more recent concerns about, say, bullying. However, given the implied acceptance of the status quo in *Barnes*, it seems likely that criminal law will continue to treat traditional assumptions about masculinity very differently to traditional assumptions about femininity in this context. This contrast carries important implications about judicial methods for finding the normative content of law. It may always be the case that social values and expectations are read into law in a selective manner, but the extent to which modern criminal law ought to allow traditional gender assumptions to inform it, and the extent to which new sources of social norms are required, is a matter for continued critical engagement.

Conclusion

This essay has shown that when it comes to responding to harm arising from inherently risky activities, the inconsistencies in the legal approaches taken by the criminal courts may by explainable by the divergent ways in

[50] *Barnes*, para. 15.
[51] *R* v. *Bradshaw* (1878) 14 Cox CC 83, cited in *R* v. *Lee* [2005] NZCA 60, para. 176.

which social dynamics influence judgment. Is it a cruel thing to subject another to a practical joke that involves throwing them up in the air and letting them fall to the ground, or setting fire to them while they sleep, or are these things merely part of the rough and tumble of growing up and becoming an adult? Is a person who has (consensual) intercourse with another without disclosing that they are HIV positive necessarily committing a criminal offence by infecting the other, or should a victim who knows such an encounter to be risky bear responsibility for it? These are questions that cannot be answered simply by taking a position vis-à-vis autonomy and paternalism. Instead, they require consideration of matters relating to public interest and social expectation about what sort of risk a person's consent authorises. I have argued that the courts have been guided by two quite different sets of gender discourses, resulting in a reading of reckless HIV transmission as an offence of predominantly male sexual violence against women, and rough horse-play as a predominantly manly diversion. Whether our aim in responding to these developments is to advocate consistency in the treatment of consent, or a less controversial (i.e. non-gendered) social basis for apparently universal legal norms, is an ongoing issue for the courts.

7 'Consensual' sexual activity between doctors and patients: a matter for the criminal law?

Suzanne Ost and Hazel Biggs

> Most codes of ethics governing professional client relations expressly forbid or strongly advise against ... sexual relations. Is this so because such relations are outside the proper scope of a professional relationship or because they cannot, within such a relationship, be fully consensual?[1]

Introduction

This essay offers the beginnings of an exploration of what counts as exploitative sexual activity between a doctor and patient, and of what role (if any) the criminal law should play. The issue of maintaining sexual boundaries between health care professionals and patients has become a significant matter of concern in recent years, following several high-profile scandals. In particular, the allegations against two consultant psychiatrists accused of sexualised behaviour towards their patients, which culminated in the publication of the Department of Health's *Kerr/Haslam Inquiry*,[2] and the GPs Clifford Ayling[3] and Oladapo Idowu have brought the issue to the attention of the public and the regulatory authorities.[4] Subsequently, the General Medical Council (GMC) produced specific guidance entitled *Maintaining Boundaries*, in which it is clearly emphasised that engaging in a sexual relationship with a patient will amount to a breach of a doctor's professional code of conduct.[5]

Our thanks to Dave Archard for his willingness to discuss our ideas over coffee, and to John Coggon for helping us realise the scale of the project we have set ourselves on exploitation.

1 D. Archard, *Sexual Consent* (Boulder, CO: Westview Press, 1998), p. 61.

2 Department of Health, *The Kerr/Haslam Inquiry: Full Report* (Cm 6640) (London: HMSO, 2005).

3 Department of Health, *Committee of Inquiry: Independent Investigation into How the NHS Handled Allegations About the Conduct of Clifford Ayling* (London: HMSO, 2004).

4 Idowu had sexual relationships with two of his patients, one of whom gave birth to his child. His name was erased from the Medical Register after a Fitness to Practise Panel found the allegations against him proven in November 2007.

5 General Medical Council, *Maintaining Boundaries* (London: GMC, 2006). The Hippocratic Oath also requires the health care professional to keep 'far from all

Doctors are advised that in order to maintain professional boundaries, they 'must not establish or pursue a sexual or improper relationship with a patient' or with a former patient, 'where at the time of the professional relationship the patient was vulnerable'.[6] In addition, 2008 saw the Council for Healthcare Regulatory Excellence (CHRE) publish guidance for regulators of health care professionals, to offer a basis for the advice they should provide to avoid breaches of sexual boundaries between health care professionals and their patients.[7]

The ethical and regulatory guidance takes the position that any engagement in sexual activity with a patient by a doctor is inappropriate and damaging.[8] In this essay, we first explore why this might be the case and whether sexual activity between doctor and patient is always exploitative. In response to Archard's question above, we consider whether consent can ever really freely be given in the context of this relationship. Does the imbalance of power between doctor and patient mean that the health care professional always effectively thwarts the patient's capacity to give free consent to sexual activity? Second, we assess whether criminal law should be utilised to deal with cases where sexual activity with a patient amounts to wrongful, harmful behaviour on the part of the doctor. We should emphasise that we present the issues in this essay by way of introduction to further work that we are currently undertaking on the issue of whether exploitation should be a matter for the criminal law.

Breaching sexual boundaries in the doctor–patient relationship

According to the CHRE, 'boundary transgressions occur wherever a professional relationship is turned into a sexual or sexualised encounter'.[9] In surveys conducted in the USA and the Netherlands in the 1980s and 1990s, it was reported that up to 10 per cent of doctors admitted

intentional ill-doing and all seduction, and especially from the pleasures of love with women and men' (D. M. Anderson, J. M. Patwell and K. Plaut (eds.), *Dorland's Medical Dictionary* (Philadelphia, PA: WB Saunders Company, 1988), p. 768).

6 General Medical Council, *Maintaining Boundaries*, paras. 4 and 5.

7 Council for Healthcare Regulatory Excellence, *Clear Sexual Boundaries between Healthcare Professionals and Patients: Responsibilities of Healthcare Professionals* (London: CHRE, 2008).

8 For a discussion of the arguments raised to justify a 'zero tolerance' approach to sexual activity between physicians and patients, see R. M. Cullen, 'Arguments for Zero Tolerance of Sexual Contact between Doctors and Patients' (1999) 25 *Journal of Medical Ethics*, 482; W. Spiegel, T. Colella and P. Lupton, 'Private or Intimate Relations between Doctor and Patient: Is Zero Tolerance Warranted?' (2005) 31 *Journal of Medical Ethics*, 27.

9 Council for Healthcare Regulatory Excellence, *Clear Sexual Boundaries*, p. 7.

having some kind of sexual encounter with a patient.[10] Transgressions[11] of this type have been attributed to situations where, due to the changing nature of the therapeutic relationship brought about by the recent emphasis on greater informality and the diminution of deferential attitudes in patients, doctors may be more likely to cross professional boundaries.[12] In Galletly's view, boundary crossings represent 'departures from usual practice that are not exploitative',[13] but if this boundary crossing becomes frequent or routine, it develops the potential to escalate into 'inappropriate relationships with patients – boundary *violations*',[14] which frequently are exploitative.

Significantly, because sexual activity is such private, intimate behaviour, even if the actions of the doctor make it wrongful behaviour on his/her part, it is hard for the patient to disclose. Not only did Ayling's patients find it difficult to extricate themselves from the immediate abusive behaviour that he subjected them to, the records show that they also felt unable to report it or even discuss it with others. *The Kerr/Haslam Inquiry* into the sexual misconduct of two doctors in Yorkshire in 2005 reported that '[i]n most if not all cases, the effect upon the women of the breach of trust that occurred has been devastating'.[15] The testimony of Clifford Ayling's patients graphically illustrates this phenomenon. One woman stated: 'I knew he had assaulted me ... but I thought I was alone. He was very controlling';[16] while another reported at his trial: 'I was a bit shell-shocked. His attitude was so brusque – it was so businesslike ... You didn't like to argue with him ... I'd been brought up to trust my doctor.'[17]

The Canadian case of *Norberg* v. *Wynrib*[18] provides further illustration of the phenomenon of a doctor breaching sexual boundaries with his

[10] N. Gartrell, J. Herman, S. Orlarte, M. Feldstein and R. Localio, 'Psychiatrist–Patient Sexual Contact: Results of a National Survey' (1986) 143 *American Journal of Psychiatry*, 1126; D. Wilbers, G. Veenstra, H. B. van de Weil and W. C. M. Weijmar Schultz, 'Sexual Contact in the Doctor–Patient Relationship in the Netherlands' (1992) 304 *British Medical Journal*, 1531.

[11] There is a dearth of accurate statistics concerning the prevalence of such transgressions in the UK, but the issues are usefully outlined in F. Subotsky, S. Bewley and M. Crowe (eds.), *Abuse of the Doctor–Patient Relationship* (London: Royal College of Psychiatrists, 2010).

[12] C. A. Galletly, 'Crossing Professional Boundaries in Medicine: The Slippery Slope to Patient Sexual Exploitation' (2004) 181(7) *Medical Journal of Australia*, 380; Speigel *et al.*, 'Private or Intimate Relations', 27.

[13] Galletly, 'Crossing Professional Boundaries', 380.

[14] Ibid. (emphasis added). [15] Department of Health, *The Kerr/Haslam Inquiry*, p. 4.

[16] 'How Did He Get Away with It for So Long?', *The Guardian*, 29 April 2002 (see www.guardian.co.uk/society/2002/apr/29/medicineandhealth.lifeandhealth).

[17] Ibid. [18] *Norberg* v. *Wynrib* [1992] 2 SCR 226.

patient. Laura Norberg was addicted to a painkiller called Fiorinal, only available on prescription. When she was unable to obtain more Fiorinal from the doctors she consulted, she went to Dr Wynrib, who initially gave her a prescription. After she made further requests for the drug, Dr Wynrib recognised she was addicted. Rather than offer her advice about anti-drug programmes that she could have enrolled in, he told her that if she was 'good to' him, then he would 'be good' to her. The implication of this, made clear to Norberg, was that he would supply the drug in exchange for sexual favours. Although she refused his offer, when Norberg was unable to obtain further prescriptions from other doctors, desperation caused her to return to Dr Wynrib and provide him with sexual favours on numerous occasions in order to have access to Fiorinal. She subsequently went to a drug rehabilitation centre of her own initiative and managed to overcome her addiction.[19]

What seems clear from the above is that the power imbalance between doctor and patient creates an environment that can facilitate a doctor's exploitation of his/her patient. We should explain here what we mean to convey by employing the term exploitation. Joel Feinberg has subjected the matter of exploitation to careful philosophical analysis,[20] arguing that 'Exploitation in the usual pejorative sense is the *wrongful turning to some advantage* by one party (A) of some trait or circumstances of another party (B).'[21] A similar understanding of the concept is also apparent in the analyses offered by Goodin, Wolff and Wertheimer.[22] Doctors such as Ayling and Wynrib thus exploit their patients because they utilise their position of power to wrongfully take advantage of their patients for their own gratification. When sexual exploitation occurs in a relationship of trust, such as that between doctor and patient, it is particularly likely to lead to the patient suffering a setback to interests,[23] such as psychiatric injury,[24] or

[19] Norberg brought claims against Dr Wynrib in battery for sexual assault, in negligence, and for breach of a fiduciary duty. All the judges in the Canadian Supreme Court awarded judgment in her favour.

[20] J. Feinberg, *The Moral Limits of the Criminal Law: Harmless Wrongdoing* (Oxford University Press, 1988), Chapters 31 and 32.

[21] Ibid., p. 192 (emphasis added).

[22] R. E. Goodin, 'Exploiting a Situation and Exploiting a Person', in A. Reeve (ed.), *Modern Theories of Exploitation* (London: Sage, 1987), pp. 166, 182; J. Wolff, 'Marx and Exploitation' (1999) 3 *The Journal of Ethics*, 105, 110–11; A. Wertheimer, *Exploitation* (Princeton University Press, 1996), p. 16.

[23] We are utilising Feinberg's understanding of harm here as a setback to an individual's interests, defined as 'things in which one has a stake', things that are vital to an individual's wellbeing. See J. Feinberg, *The Moral Limits of the Criminal Law: Harm to Others* (Oxford University Press, 1984), p. 34.

[24] See T. Allen, 'Civil Liability for Sexual Exploitation in Professional Relationships' (1996) 59(1) *Modern Law Review* 56, 69–70.

mental suffering. And patients may feel shame or guilt after the event. Laura Norberg gave testimony at the initial trial that she felt humiliation, a loss of dignity and much shame as a consequence of the sexual activities that had taken place between herself and Dr Wynrib. Moreover, the impact of any consequential harm caused by the exploitation is made worse because it amounts to a breach of the fundamental element of trust in this professional relationship. The wrongfulness and harm caused by a breach of trust and taking advantage of a relationship of power is exacerbated by the sexual character of the exploitation. As such, it constitutes particularly serious wrongful exploitation, and, in Archard's words:

> This is surely because sexuality is an area in which it is particularly important to treat one another as ends, that is to take into account and respect a person's wishes and beliefs. We are acutely aware of the dangers of failing to do so, and to that extent regard full, knowing, and considered consent as a prerequisite or at least ideal requirement of intimacy.[25]

Cases such as that of Ayling and Wynrib, where a doctor's breach of sexual boundaries involves engaging in sexual activity with the patient without consent, or using coercion or inducement, are surely cases in which it is apparent that a doctor has wrongfully taken advantage of his patient for his own gratification. What seems to make the exploitative behaviour wrongful here is the deliberate use of the more powerful position to make the patient feel that she cannot refuse the doctor's advances, or to impede the patient's choice so that she considers consent to be her only option.

But what of cases where there is no inducement on the part of the doctor, and the patient consents to the act without any enticement related to the doctor's position? It is necessary to consider whether doctors are exploiting patients in such cases, and whether a patient can ever give valid, free consent to sexual activity with her/his doctor.

Can sexual activity between doctor and patient ever be truly consensual?

Here, we consider cases where there is no coercion or inducement and the patient consents to the sexual activity. In other words, if exploitation does exist in this scenario, then it exists purely because of the imbalance of power in the doctor–patient relationship – this imbalance means, in itself, that the medical professional is wrongfully taking

[25] D. Archard, 'Exploited Consent' (1994) 25 *Journal of Social Philosophy*, 92, 99.

advantage of his/her patient for his/her own gratification. Does the imbalance of power automatically mean that the health care professional thwarts the patient's capacity to give free consent to sexual activity? In a broader context, Allen contends that it does: 'In relationships characterised by power and dependency, the relationship with the defendant may become so important to the victim that it becomes difficult to express or even make an independent choice.'[26] In a paper quoted by La Forest J in *Norberg*,[27] Phyllis Coleman argues in a similar vein to Allen, that consent to a sexual relationship in 'power dependency' relationships, such as those between health care professionals and patients, is always questionable. Notably, however, she states that: 'Exploitation occurs when the "powerful" person abuses the position of authority by *inducing* the "dependent" person into a sexual relationship, thereby causing harm.'[28] Coleman's use of the word *inducing* is significant. Indeed, we would question whether we always have an instance of exploitation (following Feinberg's explanation of the term) where consensual sexual activity between doctor and patient takes place, with no inducement. It is true that there may be cases where a doctor is able to take advantage of a patient because of the fact that she is his patient.[29] Say, for example, she undresses for an examination and he takes this opportunity to engage in sexual activity with her, and although she freely consents, the opportunity would not have existed but for the fact she is his patient. But instances of non-exploitative sexual activity between health care professional and patient, where the patient engages in consensual sexual activity without coercion or inducement, may, rarely, occur, and occasionally doctors and patients do strike up genuine relationships of affection. In these circumstances, should we perceive that the doctor has still taken advantage of the patient, merely because of the pre-existing doctor–patient relationship?

It has been argued that consensual sexual activity between a doctor and patient can *never* occur: 'Patients are not in a position to give valid consent to sexual involvement with health professionals as they are

[26] Allen, 'Civil liability', 58. See also K.R. Daniels, C. Spittal and G.B. Duff, 'Relationships between Doctors and Patients in Obstetrics and Gynaecology' (1995) 35(3) *Australian and New Zealand Journal of Obstetrics and Gynaecology*, 286, 287.

[27] *Norberg*, at para. 40.

[28] P. Coleman, 'Sex in Power Dependency Relationships: Taking Unfair Advantage of the "Fair" Sex' (1988) 53 *Albany Law Review*, 95, 96 (emphasis added).

[29] We might, for example, have a scenario that fits Archard's explanation of exploited consent: 'consent is only, and thus probably wrongly, given because of the unequal nature of the relationship between two people ... a patient [sleeps] with her therapist because he is her ... therapist' (Archard, 'Exploited Consent', 95).

bound into the unequal relationship in which they have real need.'[30] Similarly, in a case heard by the Ontario Superior Court of Justice in 2003, it was held that it was not possible for a sexual relationship between physician and patient to be truly consensual.[31] The physician in question had been found guilty of professional misconduct after the Discipline Committee of the College of Physicians and Surgeons of Ontario found that his consensual relationship with a female patient amounted to patient sexual abuse, because of the imbalance of power in the relationship.[32] In a pamphlet that the Medical Council of New Zealand sent out to registered medical practitioners in the 1990s, sexual impropriety, sexual transgressions and sexual violations that occur between a physician and a patient were defined as sexual abuse.[33] The Canadian Task Force on Sexual Abuse of Patients classifies sexual behaviour between a physician and patient into similar categories.[34] It is thus interesting to note that in a New Zealand study reported in 1994, GPs did not necessarily concur that all such behaviour amounted to sexual abuse, unless physicians made use of their more powerful position for their own gratification.[35] Yet participants still considered sexual relationships with patients to be unhealthy because of the GPs' more powerful position, and because they could consciously or unconsciously use their knowledge of the patient for their own purposes.[36]

It has also been argued that any sexual relationship between a male physician and a female patient can never be truly equal because of the inequality of power in the doctor–patient relationship, coupled with the sexual imbalance of power. In the context of such a relationship, Daniels *et al.* contend that: 'Under no professional circumstances can a woman

[30] M. Halter, H. Brown and J. Stone, *Sexual Boundary Violations by Health Professionals: An Overview of the Published Empirical Literature* (London: CHRE, 2007), p. 7.

[31] *Dr Anil Mussani* v. *The College of Physicians and Surgeons of Ontario* [2003] OJ No. 1956. See L. Ferris, 'Patient Protection Laws and the Issue of Consensual Sexual Relationships with Physicians' (2004) 170(1) *Canadian Medical Association Journal*, 34; K. H. Hall, 'Sexualization of the Doctor–Patient Relationship: Is it *Ever* Ethically Permissible?' (2001) 18(5) *Family Practice*, 511, 513 and 514.

[32] See Ferris, 'Patient Protection Laws', 34.

[33] Medical Council of New Zealand, 'Trust in the Doctor–Patient Relationship' (Wellington: Medical Council of New Zealand, 1994). See now Medical Council of New Zealand, *Sexual Boundaries in the Doctor–Patient Relationship: A Resource for Doctors* (Wellington: Medical Council of New Zealand, 2004), paras. 7–10.

[34] College of Physicians and Surgeons of Ontario, *Final Report of the Task Force on Sexual Abuse of Patients* (Toronto: College of Physicians and Surgeons, 1991). According to the Task Force: 'Sexual activity between a patient and a doctor always represents sexual abuse' (pp. 259–60).

[35] G. Eyres White, J. A. Coverdale and A. N. Thomson, 'Can One Be a Good Doctor and Have a Sexual Relationship with One's Patient?' (1994) 11(4) *Family Practice*, 389, 391.

[36] Ibid., 390.

be an equal consenting partner in the relationship. This causes women to feel abused and to suffer psychologically.'[37] The authors do not, however, provide any evidence to substantiate either claim. Without more, it is questionable why women would be considered to be especially disadvantaged, particularly as exploitation could occur in situations where the doctor is female and the patient is male, or in same-sex scenarios. However, in the aforementioned New Zealand study, when questioned about sexual relationships between GPs and patients, one participant expressed the following concerns: 'I think it is the power issue that is the important one, doctors have the power and men have the power.'[38] The authors of the study highlight the power of a medical profession, which is predominantly male, and emphasise the vulnerability of a 'predominantly female patient population', although they note that this power and vulnerability is assumptive.[39]

Today there is surely room to argue that the categorisation of the female patient population as vulnerable is an overgeneralisation. Female patients may be vulnerable to sexual exploitation in the context of some fields in medical practice, in particular, psychiatry,[40] for instance. And according to a literature review conducted for the CHRE, the majority of reported sexual boundaries breaches by a doctor involve a male health practitioner and a female patient.[41] However, not *all* female patients will be vulnerable. And undoubtedly, some patients who are vulnerable will be male, and female physicians may be more vulnerable to sexual harassment by their male patients, notwithstanding the imbalance of power in the doctor–patient relationship.[42] Schneider and Phillips note the opportunities that can exist for patients to sexualise contact with the physician: 'the nature of the doctor–patient relationship involves some unique situations. Patients are often partially nude, genital examinations are sometimes necessary, and patients may justifiably raise sexual matters in conversation.'[43]

[37] Daniels *et al.*, 'Relationships between Doctors and Patients', 287.

[38] Eyres White *et al.*, 'Can One Be a Good Doctor?', 390.

[39] Ibid., 391–2. On the vulnerability of patients generally, see J. Sheather, 'Sexual Relationships between Doctors and Former Patients' (2006) 333 *British Medical Journal*, 1132.

[40] On this, see, for example, J. Bouhoustos, J. Holroyd and H. Lerman, 'Sexual Intimacy Between Psychotherapists and Patients' (1983) 14 *Professional Psychology: Research and Practice*, 185; T. Fahy and N. Fisher, 'Sexual Contact Between Doctors and Patients' (1992) 304 *British Medical Journal*, 1519.

[41] Halter *et al.*, *Sexual Boundary Violations*, p. 7.

[42] See S. P. Phillips and M. S. Schneider, 'Sexual Harassment of Female Doctors by Patients' (1993) 329(26) *New England Journal of Medicine*, 1936; M. S. Schneider and S. P. Phillips, 'A Qualitative Study of Sexual Harassment of Female Doctors by Patients' (1997) 45(5) *Social Science and Medicine*, 669.

[43] Schneider and Phillips, 'A Qualitative Study', 675.

Can we legitimately argue that a physician who engages in sexual activity with a patient *always* exploits that patient? To reiterate, following Feinberg, we see exploitation as wrongfully taking advantage of another's weaker position for one's own benefit. However, although there will be some cases where the imbalance of power in the physician–patient relationship is exacerbated by a sexual imbalance of power and where the patient is vulnerable, we do not consider that sexual activity between doctor and patient can never be consensual. Nor do we see the imbalance of power between physician and patient as preventing the patient from being the sexual harasser. The question remains: in a situation where the sexual activity is genuinely consensual, can we say that the health care professional is wrongfully taking advantage of the patient? Whilst it may be that the doctor–patient relationship was the necessary precondition for the sexual activity, because, for example, the individuals would not otherwise have met, does the existence of this relationship necessarily mean that the patient is exploited when consensual sexual activity occurs? To establish that the activity was genuinely consensual, it must first be shown that consent was freely given and that the patient agreed without manipulation, deception or undue influence of any kind. In other words, the doctor has not set out to use his more powerful position to coerce the patient to agree to sexual activity. In the absence of inducement and/or coercion, we argue that it is much harder to demonstrate exploitation.

We do not deny that there is an imbalance of power in the relationship, given the physician's medical knowledge, awareness of the patient's medical information, and authority to provide access to medical products and treatment. However, there are grounds to suggest that there has been a reduction in this imbalance in more recent times, during which patient autonomy has taken precedence and the medical information available to patients has greatly increased. Spiegel *et al.* have argued that this is an age in which the more paternalistic relationship model between physician and patient has 'rapidly giv[en] way to informative, interpretative, or deliberative models'.[44] In this environment, criminalising consensual sexual activity without any inducement on the part of the health care professional seems to us to be wholly inappropriate, and would paint the event in an inaccurate light. Moreover, such an approach is surely inherently paternalistic towards the patient. And it does not seem to account for the possibility that we have already highlighted that physicians may also be vulnerable to sexual harassment

[44] Spiegel *et al.*, 'Private or Intimate Relations', 27.

by their patients, especially given the more recent reduction in the power imbalance. Indeed, the subject of sexual boundaries in the context of the doctor–patient relationship may generally be best approached by ensuring that both health care professionals *and* patients are aware of their responsibilities, and of the desirability of maintaining certain boundaries in their relationship.[45] Encouraging such consultation might go at least some way towards reducing the imbalance of power in the doctor–patient relationship and might help to avoid cases where exploitation does occur.

On the question of whether the patient is harmed by sexual involvement with a physician, Eyres White *el al.* note that there is no evidence to indicate that the harm suffered when a consensual, non-exploitative sexual relationship fails with a physician is any different to the harm suffered in other sexual relationships.[46] That being so, if there has been no exploitation on the part of the physician, legal intervention to address the physician's actions can be no more warranted than legal interference with the actions of parties in failed sexual relationships outside the health care context. Whilst it has been argued elsewhere that there are negative long-term emotional consequences for patients who have been involved sexually with their physicians, and these consequences have been well documented in the existing literature,[47] the majority of papers that examine such relationships relate specifically to the rather more concerning psychiatrist/therapist–patient relationship, where the situation is further complicated by the phenomenon known as transference.[48]

[45] On the ethical duties of patients generally, see M. Brazier, 'Do No Harm: Do Patients Have Responsibilities Too?' (2006) 65 *Cambridge Law Journal*, 397; H. Draper and T. Sorell, 'Patients' Responsibilities in Medical Ethics' (2002) *Bioethics* 335; Eyres White *et al.*, 'Can One Be a Good Doctor?', 392. It is clear from the existing research that patients can be the initiators of (unwanted) sexual behaviour. See Phillips and Schneider, 'Sexual Harassment' (77 per cent of participants in the survey had experienced some form of sexual harassment by a patient); Schneider and Phillips, 'A Qualitative Study', 673; J. F. Morgan and S. Porter, 'Sexual Harassment of Psychiatric Trainees: Experiences and Attitudes' (1999) 75 *Postgraduate Medical Journal*, 410 (75 per cent of the participants in the authors' survey had experienced unwanted sexual contact from patients); J. McComas, C. Hébert, C. Glacoomin, D. Kaplan and C. Dulberg, 'Experiences of Student and Practising Physician Therapists with Inappropriate Patient Sexual Behaviour' (1993) 73(11) *Physical Therapy*, 762 (80.8 per cent of participants had experienced inappropriate patient sexual behaviour).

[46] Eyres White *et al.*, 'Can One Be a Good Doctor?', 391.

[47] Hall, 'Sexualization of the Doctor–Patient Relationship', 512.

[48] See also on this point, G. A. Golden and M. Brennan, 'Managing Erotic Feelings in the Physician–Patient Relationship' (1995) 153(9) *Canadian Medical Association Journal*, 1241. We are not focusing here on sexual relationships in the context of the therapist–patient relationship, and we note that the transference phenomenon poses serious concerns regarding the validity of a patient's consent to sexual activity in this

These arguments may still fail to convince some that the existence of the doctor–patient relationship does not necessarily cause the patient to be exploited when consensual sexual activity occurs. However, we submit that when it comes to the question of legal intervention, the cases that present the strongest justification for involving the criminal law to address a doctor's violation of the sexual boundaries are those in which the doctor's breach involves her/him engaging in sexual activity with the patient without consent, or using deception, coercion or inducement. These are clearly instances of exploitation. The physician's intention will, of course, be a crucial issue. Even if there is doubt about the existence or degree of exploitation at the time the relationship begins, it may subsequently become clear that the intention of the doctor was always only to exploit the patient for sex, hence intention is/may be an important factor in determining whether the encounter was exploitative, a matter we shall explore in our further work. But the question that now needs to be answered is when will or should the criminal law intervene in such cases?

Sexual exploitation in the doctor–patient relationship and the limits of the criminal law

It is the question of whether consent has been given to sexual activity that frames the applicability of the criminal law to a doctor's violation of the sexual boundaries with his/her patient. Sexual activity that occurs between a doctor and patient without the patient's valid consent is criminalised by the offences of rape, assault by penetration, sexual assault or causing sexual activity under the Sexual Offences Act 2003 (SOA).[49] An offence is also committed under the SOA where Y and X are in a relationship of care and where X has a mental disorder that impedes choice, if Y sexually touches X, engages in sexual activity in the presence of X, causes or incites sexual activity by X or causes X to watch sexual activity.[50] If Y is a health care professional working for the NHS or offering private medical care that involves providing services to X, then Y is in a relationship of care with X for the purposes of these offences.[51] In addition, sexual activity and behaving in certain sexual ways with children is criminalised under sections 16–19 when the adult is in a position of trust with the child. Doctor–patient relationships are included here.[52]

relationship. On this, see, for example, Bouhoustos *et al.*, 'Sexual Intimacy'; Fahy and Fisher, 'Sexual Contact'.

49 Sexual Offences Act 2003, sections 1–4.
50 Ibid., sections 38–41.
51 Ibid., section 42(3).
52 Ibid., section 21(4).

But whether there would be any criminal law action in other cases where the patient consents, but does so because the doctor's more powerful position induces, entices or causes her to be unable to say no, is more questionable. Consent remains a nebulous concept in the context of sexual activity under the SOA. Section 74 states that 'a person consents if he agrees by choice, and has the freedom and capacity to make that choice'. At issue is whether patients such as Laura Norberg, who consent because of some inducement or coercion on the part of the doctor, can be shown to have agreed by choice. None of the evidential or conclusive presumptions that establish non-consent under section 75 of the SOA would apply here.[53] Although we could question whether patients who are induced, enticed or unable to say no to sexual activity with their doctors have consented freely, they fall outside the parameters of the presumptions. This means that the issue of whether they consented would be left to the jury, applying the general definition of consent under section 74.[54] Here the difficulty would lie in assessing whether or not the patients are 'agreeing by choice' and have 'the freedom and capacity to make that choice', without any further guidance as to how this assessment is to be made.

Critical of the SOA's definition of consent, Ashworth comments that, 'freedom of decision-making may be greater or less, depending on the impact of any deception, threats or other perceived pressures, and the question is what degree of impairment should be taken to mean that any apparent consent was not free'.[55] In the view of Elliott and de Than, 'The section 74 statutory definition places an emphasis on personal autonomy, but the use of ambiguous concepts of freedom and choice

[53] The evidential presumptions under section 75 are as follows: if violence is used or the complainant is caused to fear that immediate violence would be used against the complainant or another person; if the complainant was, but the defendant was not, unlawfully detained at the time; if the complainant was asleep or unconscious at the time; because of the complainant's physical disability, the complainant would not have been able to communicate to the defendant whether he consented; anyone had administered to or caused to be taken by the complainant, without the complainant's consent, a substance that was capable of causing or enabling the complainant to be stupefied or overpowered at the time. The conclusive presumptions under Section 76 apply where the defendant intentionally deceived the complainant as to the nature or purpose of the act; the defendant intentionally induced the complainant to consent by impersonating a person known personally to the complainant.

[54] See J. Temkin and A. Ashworth, 'The Sexual Offences Act 2003: (1) Rape, Sexual Assaults and the Problems of Consent' [2004] *Criminal Law Review* 328, 336.

[55] A. Ashworth, *Principles of Criminal Law* (6th edn.) (Oxford University Press, 2009), p. 339.

leaves the definition extremely vague.'[56] And case law has not, as yet, offered any real guidance other than in specific contexts.[57]

Therefore, the limits of the criminal law's ambit in cases of a breach of the sexual boundaries by a doctor are determined by the matter of consent, and the matter of what amounts to impaired consent remains a grey area. Framing the criminal law around the matter of consent in the context of sexual exploitation is open to the criticism that the law is only emphasising one particular wrong of such behaviour and not drawing attention to other possible wrongs, such as the abuse of trust. Indeed, it may be that the wrongs of sexual exploitation in the doctor–patient relationship would only be fully taken into account by the criminal law if there were a different framework in place. Perhaps, then, this framework should revolve around exploitation in a relationship of trust, as is the case regarding other SOA offences, such as those previously mentioned involving sexual activity and behaving in certain sexual ways with children when the adult is in a position of trust with the child, and where Y and X are in a relationship of care and X has a mental disorder that impedes choice.[58]

In order to extend the criminal law's ambit as we are suggesting, it is of course necessary to consider whether utilising the criminal law to deal with breaches of the sexual boundaries by a doctor in cases where a patient is induced or coerced into consenting by creating a new offence is appropriate. As noted by Ashworth, 'a core element of criminal law, from a normative point of view, is that the criminal sanction should be reserved for substantial wrongdoing. And this prompts questions about how the seriousness of wrongdoing should be assessed.'[59] Our

[56] C. Elliott and C. de Than, 'The Case for a Rational Reconstruction of Consent in Criminal Law' (2007) 70(2) *Modern Law Review*, 225, 238. Similarly, Dennis has commented that 'broad notions of "choice" and "freedom" [within section 74] leave a good many questions unanswered about the kinds of non-violent threats or other pressures that might invalidate an apparent consent' (I. Dennis, 'The Sexual Offences Act' [2004] *Criminal Law Review* 79).

[57] For example, the Court of Appeal decision in *R* v. *Jheeta* [2007] EWCA Crim 1699 suggests that pressure the defendant placed on the complainant to have intercourse with him by a 'complicated' and 'unpleasant' scheme he had fabricated meant that she did not consent, but as Elvin comments, 'it says nothing at all about other types and degrees of pressure'. J. Elvin, 'The Concept of Consent under the Sexual Offences Act 2003' (2008) 72 *Journal of Criminal Law*, 519, 529. The element of the defendant's deceit as to the reality of the situation may have played a role in the judgment (see para. 28).

[58] The concept of exploitation has been utilised as a descriptor for these offences. See, for example, Sentencing Advisory Panel, 'The Sexual Offences Act 2003 – The Exploitation Offences: A Consultation' (2004) 168(17) *Justice of the Peace & Local Government Law*, 308.

[59] A. Ashworth, 'Is the Criminal Law a Lost Cause?' (2000) 116 *Law Quarterly Review*, 225, 240.

assessment of how serious sexual exploitation in the doctor–patient relationship is revolves around the nature and effects of this behaviour, which we have presented above. As the *Norberg* case demonstrates, such exploitation may give rise to a more than trivial risk of causing victims to suffer significant psychological or physical harm and can therefore be seen to be *prima facie* wrongful.[60] According to the literature review conducted for the CHRE, '[t]he impact on survivors of professional sexual boundary violations/abuse shows considerable and enduring harm', with symptoms including 'post traumatic stress disorder, anger, a sense of betrayal and exploitation, guilt and self-blame'.[61] Doctors owe an ethical duty to their patients not to cause them harm, and the breach of the trust placed in them by their patients, and, indeed, the public, when they sexually exploit their patients, is both serious and damaging. Thus, we argue that a doctor's sexual exploitation of his/her patient involving coercion or inducement to engage in sexual activity can constitute a public wrong, since it contravenes defining values (such as trust and protection of the less powerful in societal relationships) that society recognises as fundamental and that the state should endeavour to safeguard to ensure the good of its citizens.[62]

There might be less of a case for involving the criminal law if the civil law provided a remedy. However, a civil law action under the intentional torts would again involve the problematic issue of consent.[63] If the doctor–patient relationship were to be viewed through the lens of fiduciary duties in this jurisdiction, this might well offer an alternative to invoking the criminal law to deal with the sexual exploitation we have discussed.[64] But finding fiduciary liability on the basis of a breach of trust remains a path that our courts do not seem prepared to travel down in the context of the doctor–patient relationship.[65] Moreover, even

60 M. Madden Dempsey and J. Herring, 'Why Sexual Penetration Requires Justification' (2007) 27(3) *Oxford Journal of Legal Studies*, 467, 479.

61 Halter *et al.*, *Sexual Boundary Violations*, p. 7.

62 On the matter of public wrongs, see S. E. Marshall and R. A. Duff, 'Criminalization and Shared Wrongs' (1998) XI(1) *Canadian Journal of Law and Jurisprudence*, 7, 20.

63 Whilst there is authority to suggest that a doctor's position of power may cause consent to be less than real, this is in the specific context of a prison setting. See *Freeman* v. *Home Office* [1984] 1 All ER 1036.

64 In *Norberg* v. *Wynrib*, two of the judges in the Supreme Court found that Dr Wynrib owed, and had breached, a fiduciary obligation to his patient. See, especially, McLachlin J's judgment, at paras. 61–121.

65 See Lord Scarman in *Sidaway* v. *Board of Governors of the Bethlem Royal Hospital and the Maudsley Hospital* [1985] 1 All ER 643, at 651. See also M. Brazier and M. Lobjoit, 'Fiduciary Relationship: An Ethical Approach and a Legal Concept?' in R. Bennett and C. A. Erin (eds.), *HIV and AIDS Testing, Screening and Confidentiality* (Oxford University Press, 1999), p. 179.

if the civil law could be invoked in this way, we would still contend that such exploitation is a public wrong that warrants the attention of the criminal law.

Conclusion

There is no doubt that health care professionals should not pursue or engage in sexual relationships with their patients, and that they will be brought to account via professional regulation if they do so and their behaviour is discovered. Notwithstanding this, we have argued that a genuinely consensual relationship occurring between a doctor and patient does not involve the wrongful taking advantage of the patient for the doctor's own ends and thus does not constitute exploitation. But we submit that such genuinely consensual sexual activity is rare.

Given the inherent power imbalance and breach of trust, the validity of the patient's consent can often be questioned, and, in our view, exploitation does occur when the doctor obtains the consent through deception, inducement or coercion. We have suggested that such sexual exploitation should rightly be a matter for the criminal law. But although the criminal law on sexual offences captures some of this exploitation, there is an important impairment of consent issue that is not tackled by the evidential and conclusive presumptions that establish non-consent to sexual acts under the SOA. Because whether or not the patient has given free consent is thus a matter for the jury (applying a loose and ambiguous statutory definition of what counts as valid consent), it is not clear whether a doctor would be considered to have committed an offence under the SOA if s/he obtains the patient's consent through coercion or inducement. Moreover, even if such behaviour is caught by existing sexual offences through a jury's interpretation of the consent provision under the SOA, the existing law does not spell out what makes this behaviour especially wrongful: namely, exploitation in the context of a relationship of trust.

One way in which this could be addressed, at least to some degree, is to introduce a new abuse of trust offence. This could apply where a medical professional violates the sexual boundaries with a patient and acts in a way that impedes the patient's choice to engage in sexual behaviour. An analysis of what the specific elements of and terminology within this offence should be, and whether such an offence should also apply in the context of other relationships of trust and unequal power, is a matter for another paper. But, given our contention that criminalisation of sexual exploitation can be appropriate in the specific context of a relationship involving trust and unequal power,[66] it is significant to note that the wrongs and harms that such behaviour can cause may be caught by the criminal law in another jurisdiction. Notably, there is no consent, for the purposes of the offence of sexual assault under the Canadian Criminal Code, where 'the accused induces the complainant to engage in the activity by abusing a position of trust, power or authority'.[67] This invocation of the criminal law to deal with such sexual exploitation provides much food for thought.

[66] And further work will follow, exploring in greater detail what we consider should count as exploitation that warrants criminal law's intervention.
[67] Canadian Criminal Code, section 273(1)(2)(c).

Part III

Criminalising biomedical science

8 'Scientists in the dock': regulating science

Amel Alghrani and Sarah Chan

> Occasionally science makes procedures possible that are so radical that those at the interface between science and society are called on to define moral standards for society.[1]

Introduction

The relationship between science and society is at times an uneasy one. On one hand, although the application of the modern scientific method is a relatively recent development, humans since the beginning of history have always sought to understand the world around us and to use that understanding to improve the human condition. Some might even argue that our curiosity and our desire to satisfy it through scientific inquiry are defining characteristics of what it means to be human. Certainly, science and its applications have provided tremendous benefits to humankind – for example, in terms of improvements to health and welfare. At the same time, however, science presents us with challenging social and ethical difficulties as the progress of technology opens up greater possibilities for changing and controlling our environment and even ourselves.

Keeping pace with fast-moving science and the ethical controversies to which such research may give rise has often presented a challenge to regulators.[2] Public concerns over science must be allayed, whilst also ensuring scientists submit to standards of social acceptability, or at least ethical oversight. Yet the government must also consider the ideal of scientific freedom, which holds that the pure pursuit of knowledge, if it

[1] M. Warnock, 'The Ethical Regulation of Science' (2007) 450(20) *Nature*, 615.

[2] Take, for example, the issue of reproductive cloning: it was only when Dolly the sheep attracted media attention and a subsequent challenge was brought regarding the legality of human reproductive cloning (*R (Quintavalle)* v. *Secretary of State for Health* [2001] EWHC 918 (Admin)) that the government reacted by rushing the Human Reproductive Cloning Act 2001 through Parliament. The issue was dealt with more fully in the Human Fertilisation and Embryology Act 2008.

is to achieve its full potential, cannot be unduly influenced by social norms, nor fettered by constraints imposed by society. This dualistic, adversarial view of science and society as competing entities is not entirely apposite, given the mutuality of their natures: public trust, communication and engagement also play a vital role. In this complex environment, therefore, the question of how science is regulated has assumed an increasing importance as society seeks to realise the benefits of research, whilst keeping the goals and means of science aligned with ethically acceptable norms.

This essay endeavours to provide an exploration and analysis of the use of the criminal law to regulate science. First, we consider the putative need for regulation of science and the appropriateness of imposing external legal regulations versus reliance on internal institutions of science, such as scientific responsibility and self-regulation. Second, we examine the use of criminal law in regulating two particular areas of science: research involving reproductive biomaterials and research involving human tissue and transplantation. In the final part of this essay, we consider general concerns underlying the use of criminal law to regulate science and its role in this context.

Regulating science: how and by whom?

Scientific responsibility: moral code, code of conduct?

> Scientists are the gatekeepers of new knowledge and as such they have a special responsibility to the rest of society.[3]

The requirement for scientific responsibility often found in self-directed guidelines issued by various professional scientific organisations is a key part of the science–society relationship. As Verhoog puts it, '[a] large measure of freedom is essential to the pursuit of science. This freedom should also extend to the exercise of a correspondingly large measure of social responsibility by the scientists themselves.'[4] Scientific responsibility thus defines a sort of moral code for science, and perhaps a form of regulation or control, insofar as it also constitutes a code of conduct.

Scientists evidently have responsibilities with respect to the conduct of research: to observe the requirements of scientific method to ensure the veracity of results; to report these results truthfully and accurately; and

[3] International Council for Science, *Freedom, Responsibility and Universality of Science* (Paris: ICSU, 2008), p. 13 (see www.icsu.org/publications/cfrs/freedom-responsibility-booklet).

[4] H. Verhoog, 'The Responsibilities of Scientists' (1981) 19(4) *Minerva*, 582, 603.

to conduct their research as effectively and efficiently as possible. Such responsibilities are operationally required, quite apart from any moral imperatives that may also apply, to maintain science itself. If falsification of data were standard practice, for example, the internal reliability and cohesion of scientific knowledge would break down, rendering the entire practice of science as we understand it pointless. (The results of a recent study reporting an alarming incidence of fabrication, falsification and other questionable research practices are all the more disturbing in light of this analysis.[5])

As well as these internal responsibilities, scientists have external responsibilities relating to the role of science in society. The obligations of scientists to society, and vice versa, have been described in terms of a social contract: 'an arrangement built on trust which sets out the expectations of the one held by the other, and which – in principle – includes appropriate sanctions if these expectations are not met'.[6] But what does this actually entail?

A key issue is whether scientists should bear responsibility for potential uses of their research findings, particularly applications that may lead to harmful consequences. The general moral principle of harm avoidance as applied to science raises difficulties with respect to research that can produce both great potential benefit and great harm – a problem known as the dual-use dilemma.[7] The paradigm case of this, of course, is nuclear physics research and the development of the atomic bomb; in recent times, much concern has arisen over biological weapons technology and the types of research that might pave the way for development of bioweapons. Legally speaking, the regulatory response so far has been to focus on the development of harmful applications rather than the pursuit of 'basic' research;[8] but in terms of moral responsibility, and as the line between basic and applied research becomes ever more blurred and less relevant, this becomes a spurious distinction.[9]

Scientists also bear responsibility for communicating their research, not only accurately and truthfully, but with regard to the social implications

[5] D. Fanelli, 'How Many Scientists Fabricate and Falsify Research? A Systematic Review and Meta-Analysis of Survey Data' (2009) 4(5) *PLoS ONE*: e5738. doi:10.1371/journal.pone.0005738.

[6] M. Gibbons, 'Science's New Social Contract with Society' (1999) 402(6761) *Nature*, C81–4.

[7] H. J. Ehni, 'Dual Use and the Ethical Responsibility of Scientists' (2008) 56(3) *Archivum Immunologiae Therapiae Experimentalis*, 147.

[8] For example, through the Bioweapons Act 1974.

[9] For example, research into bioweapons development is closely regulated, but basic microbiological research that is a prerequisite for such work and might generate knowledge leading directly to these applications, is not controlled in the same way.

of their work. In addition, it may be argued, they have a broader responsibility to communicate accurately about scientific matters in general, in view of how they are likely to be received by the lay public and the trust placed in scientists as representatives of science as a whole. Whilst not necessarily justifying a more stringent legal standard for scientific representation, this element of trust, we shall argue, may be an important factor in the use of criminal law to regulate research, in that the presence of criminal sanctions, even if seldom invoked, may provide public reassurance and hence social legitimisation in controversial areas.

One might argue that scientists also have a responsibility to use science for the public good, as a specific manifestation of the duty of beneficence: as agents who have control of scientific knowledge, scientists are best equipped to use that knowledge to benefit others. An interesting issue, though beyond the scope of this essay, is how this putative obligation relates to the commercialisation of science and private control of research, which may take place in ways that are not conducive or even downright incompatible with the greatest good. We may note, however, that the harms (or failures to benefit) inflicted by limitations on access to science and its fruits, such as restrictions on availability of health care products and development technologies, might well exceed the harms that are likely to be inflicted directly by most forms of research. This being the case, it might seem paradoxical for so much concern for the regulation of science to be directed proscriptively at the conduct of individual scientists, rather than prescriptively as to how science (and its supporters, commercialisers and 'owners') ought to be using it for good – but this is a topic for another discussion.

Self-regulation: is it sufficient?

Scientific responsibility provides a framework for assessing the conduct of scientists, but can it also function in a normative manner to guide or control that conduct? It has been argued that the burden of scientific responsibility 'cannot be borne alone by individual scientists ... Part of this responsibility certainly has to be assumed by institutions created by the scientific community itself and by their recommendations.'[10] The scientific community and its recommendations should, of course, have input into the regulation of science. The question is whether 'institutions created by the scientific community' are sufficient to regulate science or whether further external regulation by law is necessary, and, if it is, whether the criminal law is the appropriate mechanism.

[10] Ehni, 'Dual Use', 149.

Many other professions incorporate some element of self-regulation that functions additionally to or in tandem with law. For example, to draw a further comparison with medicine, in the UK the General Medical Council (GMC) is responsible for overseeing the activities of doctors.[11] Other professions operate various schemes: some as sole national self-regulators, others as one of a number of self-regulation organisations for the profession; some with penalties for non-compliance, such as withdrawal of licences or disaffiliation; some reinforced by legal provisions, others not.

There are arguments both for and against self-regulation of science. In favour of self-regulation, it might be argued that scientists are best placed to understand their work and its implications, and therefore how it ought to be regulated. If regulation is left up to legislators and policy-makers who do not fully understand the science, it risks being irrelevant, inappropriate and ineffective. Regulation created by a professional body of scientists might be more effective, better suited to the needs of the profession and more likely to facilitate the progress of science in an appropriately controlled manner.

Against this, a major problem with self-regulation is that currently, there is no such thing as a registration scheme for practising scientists or a 'licence to practise' science, nor a single professional governing body even within the national jurisdiction, let alone internationally. How, then, can self-regulation be enforced? Critics point to the impracticality of enforcement as one reason why self-regulation is unlikely to be effective.

This view of self-regulation as ineffective without enforcement provisions assumes what might be considered an unduly negative view of scientists: that they require the threat of punishment to deter them from 'going wrong', and that without such a threat they will run amok, committing all sorts of moral crimes. Most scientists, one hopes, are not like that. The occasional unfortunate instances of misconduct, however, diminish public faith in science and perhaps indicate the need for a level of external regulation.[12]

How does the law currently regulate science?

The regulation of science is far from a uniform area of law. No single piece of legislation exists to regulate science as a whole; while the UK

[11] The responsibilities and powers of the GMC, established by the Medical Act 1983, include operating a registration scheme for practitioners, setting professional standards and evaluating adherence to them, and enforcing disciplinary measures as required.

[12] See, for discussion, B. L. Willcox, 'Fraud in Scientific Research: The Prosecutor's Approach' (1992) 2(2) *Accountability in Research*, 139.

government has recently developed proposals for a single research regulator,[13] this applies only to 'medical research', not science in general. There are a number of organisations concerned with the maintenance of research integrity across all areas of science,[14] but in many instances their role is advisory rather than regulatory, and in any case, the remit of such organisations is generally limited to 'good research practice'.

Scientific activity is of course subject to general laws that might apply to constrain what may be done in the course of research. Thus, for example, scientists may need to have regard to legal requirements regarding the supply of research, and to occupational regulations governing the terms and conditions under which researchers work,[15] both of which may present special issues in the context of scientific research.

The question with which we are concerned, however, is whether science by its nature requires some additional level of regulation. In the UK, there are laws and regulations governing some particular types of research – for example, animal research is regulated by the Animals (Scientific Procedures) Act 1986.[16] Another example is the set of regulations relating to the use of genetically modified organisms, including a specific provision of the general-application Environment Protection Act 1990,[17] as well as various statutory instruments.[18]

Medical research poses questions of special interest, falling on the intersection between health care (regulation of which is itself controversial) and science. Research directly involving medical treatment of humans will be governed in the same manner as other medical treatment. Clinical trials involving medicines (though not other forms of clinical trial) are regulated by statute[19] in the UK; various other international conventions and protocols also govern clinical trials,[20] though these do not carry the force of law.

[13] Department of Health, 'Liberating the NHS: Report of the Arm's-Length Bodies Review' (London: Department of Health, 26 July 2010) (see www.dh.gov.uk/en/Publicationsandstatistics/Publications/PublicationsPolicyAndGuidance/DH_117691).

[14] For example, in the UK, the Research Integrity Office.

[15] For example, the Health and Safety at Work Act 1974.

[16] This Act requires that a licence be obtained for certain forms of research on animals, sets out the requirements for licensing and the terms that must be complied with, and makes provision for penalties for non-compliance, including for performing research activities (that would have been lawful if a licence were granted) without a licence.

[17] Section 108(1).

[18] The Genetically Modified Organisms (Contained Use) Regulations 2000, SI 2000/2831; Statutory Instrument 1996 No. 1106; the Genetically Modified Organisms (Deliberate Release and Risk Assessment-Amendment) Regulations 1997, Statutory Instrument 1997 No. 1900.

[19] The Medicines for Human Use (Clinical Trials) Regulations 2004.

[20] For example, the Declaration of Helsinki.

In examining the regulation of science through criminal law, we shall concentrate on UK regulation of two areas: research involving human reproductive biomaterials, and research on human tissue. These are of particular relevance because although they pertain to medical practice, the law governing them extends to research not directly involving medical treatment of patients. We consider the implications of this tandem regulation and examine the justifications for the invocation of criminal law in this context.

Research involving reproductive biomaterials

Research involving human reproductive materials is regulated by the Human Fertilisation and Embryology (HFE) Act 1990 (as amended), with oversight by the Human Fertilisation and Embryology Authority (HFEA).[21] The development of *in vitro* fertilisation (IVF) generated public 'unease at the apparently uncontrolled advance of science, bringing with it new possibilities for manipulating the early stages of human development',[22] and regulation of this area, as Deech notes, 'initially arose from the widespread fears expressed by politicians, the media, the public and the professions about embryo research, its morality and its direction'.[23] The HFE Act, passed in 1990 was, according to Montgomery, the 'first attempt in English law to provide a comprehensive framework for making medical science democratically accountable'.[24] The government held both 'assisted reproduction and research involving the embryo of the human species' to be 'legitimate interests of the state', stating that 'reproductive and research freedoms must be balanced against the interests of society'.[25] Regulation via statute with oversight by a statutory authority was said to offer certainty and protection to professionals, who would be provided with clear legal and professional guidelines, whilst simultaneously protecting patients,

[21] Although, as noted earlier, changes to the regulation of medical research have recently been announced, since there have been no government statements to indicate otherwise, it is expected that specialist regulation of this area via statute will remain in force.

[22] Department of Health and Social Security, *Report of the Committee of Inquiry into Human Fertilisation and Embryology* ('The Warnock Report'), Cmnd 9314 (London: HMSO, 1984), p. 4.

[23] R. Deech, 'The HFEA: 10 Years On' in J. Gunning and H. Szoke (eds.), *The Regulation of Assisted Reproductive Technology* (Aldershot: Ashgate, 2003), p. 21.

[24] J. Montgomery, 'Rights, Restraints and Pragmatism: The HFE Act 1990' (1991) 54 *Modern Law Review*, 524.

[25] House of Commons Science and Technology Committee, *Human Reproductive Technologies and the Law* (HC-7), Fifth Report of Session 2004–5 (London: TSO, 2005), paras. 46–47.

since doctors and scientists would be deterred from irresponsible behaviour. The government stated:

> The HFE Act was drawn up on the basis that certain activities involving human embryos outside the body, or the use of stored or donated gametes, demanded active regulation and definite legal limits.[26]

Despite the various challenges to the HFE Act since its inception in 1990,[27] in 2008 the government retained its position that research involving human reproductive materials continues to merit specialist regulation:

> Ultimately, the Government believes that the force of law remains justified in the distribution of permissions, rights, responsibilities and prohibitions for the development and use of human reproductive technologies. Law and active regulation are necessary to set out and monitor a system of public oversight and accountability, taking account of the principles of good regulation.[28]

It is not surprising that such a stance was adopted, since the complex ethical, social and political questions raised by the use of human embryos continue to attract controversy.[29]

Criminal sanctions and the HFE Act

The HFE Act 1990 provides a regulatory framework to govern research on reproductive biomaterials, and specifies that the creation, storage and use of human gametes can only be carried out under a licence granted by the HFEA, for the purpose of fertility treatment, storage and/or research.[30]

[26] Department of Health, *Impact Assessment on the Human Fertilisation and Embryology Bill (2008)* (8 November 2007), para. 7 (see www.dh.gov.uk/en/Publicationsandstatistics/Legislation/Regulatoryimpactassessment/DH_080209).

[27] Legal challenges on numerous issues, ranging from disputes over frozen embryos (*Natalie Evans* v. *Amicus Healthcare Ltd and Others* [2004] EWCA (Civ) 72, [2004] 2 FLR 766, CA), the creation of saviour siblings (*R (on the application of Quintavalle)* v. *HFEA* [2003] 3 ALL ER 257) and the controversy surrounding reproductive cloning (*R* v. *Secretary of State for Health, Ex p Quintavalle* [2003] 2 WLR 692), all exposed the weaknesses of the HFE Act 1990.

[28] Department of Health, *Impact Assessment*, para. 7.

[29] R. Gledhill, 'Public Want More Debate on Human Embryo Research', *The Times*, 13 May 2008; C. Hastings, 'Catholic Primate Calls for Debate on Embryo Research Before Fox-hunting', *The Telegraph*, 5 July 2003; A. Charo, 'The Hunting of the Shark: The Moral Status of Embryo, Right-to-Lifers, and Third World Women' (1995) 6 *Stanford Law Policy Review*, 2; M. Mulkay, *The Embryo Research Debate: Science and the Politics of Reproduction* (Cambridge University Press, 1997); R. Green, *The Human Embryo Research Debates* (New York: Oxford University Press, 2001); J. Savulescu, 'The Embryonic Stem Cell Lottery and the Cannibalization of Human Beings' (2002) 16(6) *Bioethics*, 508.

[30] Note that following amendments made to the 1990 Act by the Human Fertilisation and Embryology (Quality and Safety) Regulations (SI 2007/1522) implementing the

Licence Committees of the HFEA are dubbed 'the gatekeepers for all licensed activities in this field'.[31] Activities governed and limited by licences are detailed;[32] carrying out any of these activities unlicensed is a criminal offence and may result in a fine, a custodial sentence for up to ten years, or both.[33] Section 41 of the HFE Act 1990 lists the activities that constitute criminal offences, including: implanting non-human or admixed embryos or gametes in a woman; using female germ cells taken or derived from an embryo or foetus to provide fertility services; creating, keeping or using admixed embryos other than in pursuance of a licence; and putting human or admixed embryos in animals.

At the time of writing, criminal prosecutions in the domain of assisted reproduction have been extremely rare. The only successful prosecution brought under the HFE Act arose in 2009, when two men were convicted for running a fertility website, without a licence, selling sperm from anonymous donors to women trying to conceive.[34] They each received a nine-month custodial sentence, suspended for two years,[35] a £15,000 fine and 200 hours of unpaid work, and were banned from future work in the industry. The only other criminal prosecution related to this field was in 2002, when an embryologist was charged with and found guilty of assault causing actual bodily harm and false accounting in relation to treatments administered to patients at a private fertility clinic. Dr Paul Fielding was responsible for thawing and preparing embryos for injection into the wombs of the patients who attended the clinic for IVF treatment. Instead of preparing the embryos as expected, in several cases he substituted saline solution. Upon conviction, he received an eighteen-month custodial sentence and was ordered to compensate the victims.[36]

Despite so few successful criminal prosecutions being brought for activities within the domain of IVF, criminal breaches of licence conditions are often detected when licensing inspections are carried out, but are infrequently punished.[37] Davies and Veblen note that two criminal

European Union Tissue and Cells Directive, a licence is now also required under the 1990 Act in respect of non-medical fertility services.

31 Human Fertilisation and Embryology Authority (see www.hfea.gov.uk/140.html).

32 HFE Act 1990, Schedule 2.

33 HFE Act 1990, sections 41 and 42.

34 Since April 2007, under the Human Tissue (Quality and Safety for Human Application) Regulations, any provider wanting to 'procure, test, process or distribute' unfrozen sperm must hold a licence from the HFEA.

35 'Illegal Fertility Company Owners Given Suspended Prison Sentence', *The Guardian*, 12 October 2010.

36 See HFEA, 'Statement on the Hampshire case' (12 December 2003) (see www.hfea.gov.uk/859.html).

37 T. Davies and K. Veblen, HFEA Authority Paper, *Regulatory Enforcement and Sanctions Act 2008* (17 December 2008), para. 5.4 (see www.hfea.gov.uk/docs/AM_Item7_Dec08.pdf).

offences are commonly discovered in inspections: storage without effective consent and breaches of confidentiality, yet enforcement action is rarely taken, even when offences are repeatedly committed.[38] This attracted criticism that 'the HFEA has not yet developed consistent practices of enforcement'.[39]

This apparent inconsistency could stem from the fact that the enforcement options available to the HFEA when a criminal offence is committed are limited.[40] If so, the Regulatory Enforcement and Sanctions Act 2008 (RES Act) might have provided an opportunity for the HFEA to broaden its ability to ensure compliance, by providing 'an extended tool kit of alternative civil sanctions as a more proportionate and flexible response to cases of regulatory non-compliance normally dealt with in the criminal courts'.[41] The RES Act only applies to criminal offences, such as those section 41 HFE Act offences mentioned earlier. A preference to ensure compliance through the RES Act might reflect a reluctance to use the criminal process in this domain.

Whilst it is clear, then, that criminal law is in theory available to regulate this area of science, its provisions are seldom enforced in practice. This leads to the question of what role the criminal law actually fulfils in this context. We shall return to this later.

Research involving human tissue and organs

Research involving human tissue is another area subject to specific statutory regulation, via the Human Tissue Act 2004,[42] with oversight by the Human Tissue Authority (HTA).[43] Like the HFE Act 1990, this Act was also born amidst controversy, being (as Price describes it) 'principally a response to the furore generated by revelations about practices relating to the retention and use of human tissue in the Bristol

[38] Ibid., para. 6.3.6. [39] Ibid., para. 8.

[40] They are limited to informal actions, direct action in respect of a licence, or referral to the police.

[41] Better Regulation Executive, *Regulatory Enforcement and Sanctions Act 2008: Guidance to the Act* (London: BRE, 2008), p. 7.

[42] The 2004 Act is now the primary legislation regulating transplantation in England, Wales and Northern Ireland, but it does not apply in Scotland. For more on this statute, see D. Price, 'The Human Tissue Act 2004' (2005) 68(5) *Modern Law Review*, 798, and J. McHale, 'The Human Tissue Act 2004: Innovative Legislation – Fundamentally Flawed or Missed Opportunity?' (2005) 26 *Liverpool Law Review*, 169.

[43] Again, the functions of the HTA may soon be transferred to the Care Quality Commission, although it is expected that it will continue to be regulated via statute.

Royal Infirmary (Kennedy)[44] and Alder Hey Children's Hospital in Liverpool (Redfern) Inquiry Reports'.[45]

The statutes previously governing this area were the Human Tissue Act (1961), the Anatomy Act (1984) and the Human Organ Transplants Act (1989). The original Human Tissue Act 1961 consisted of just four sections and provided no legal remedies for breach, which Price notes was 'a quite remarkable lacuna'.[46] The Human Tissue Act 2004 sought to remedy this and repealed the former Human Tissue Act (1961), Anatomy Act (1984) and the Human Organ Transplants Act (1989).

The 2004 Act provides that in order to store and carry out research on all 'relevant material', a licence must be obtained from the HTA. 'Relevant material' includes any material that consists of human cells other than hair or nails,[47] and gametes or embryos (which are already regulated under the Human Fertilisation and Embryology Act 1990).[48] Section 1 makes reference to the need for 'appropriate consent' to be given for the use of 'relevant material' for the purposes listed in Schedule 1, including research.

Criminal sanctions in the Human Tissue Act 2004

Researchers who fall foul of the provisions contained within the 2004 Act may be liable to criminal prosecution. A person who is found to have failed to have had reasonable belief that 'appropriate consent' had been obtained for the removal, storage or use of relevant material for any of the scheduled purposes,[49] stores or uses human tissue donated for a specified purpose for a different purpose,[50] participates in the trafficking of human tissue for transplantation, or carries out licensable activities without holding a licence from the HTA,[51] is liable to a fine, up to three years' imprisonment on indictment, or both.

The criminal sanctions appear to have arisen because other remedies were deemed inadequate. Civil law failed to offer much of a remedy for those aggrieved by the unlawful retention of organs, as the injury was regarded as psychiatric in character;[52] nor did the previous statute allow

[44] Bristol Royal Infirmary Inquiry, *Interim Report: Removal and Retention of Human Material* (London: TSO, 2000).

[45] The Royal Liverpool Children's Inquiry, *Report*, HC [Session 2000–1] (London: TSO, 2001); Price, 'The Human Tissue Act', 799.

[46] Price, 'The Human Tissue Act', 799.

[47] Excluded on account of their 'natural discardability'.

[48] Relevant material is defined in the Human Tissue Act 2004, section 53.

[49] Ibid., section 5. [50] Ibid., section 8. [51] Ibid., section 25.

[52] See M. Jones, 'Retained Organs: The Legal Fallout' (2004) 20(3) *Professional Negligence*, 182.

for criminal charges to be brought. This caused much concern. For example, Professor Van Velzen, the pathologist who, from 1988 to 1995, secretly retained hundreds of dead children's organs at Alder Hey hospital, ultimately escaped conviction. The Royal Liverpool Children's Inquiry[53] reported that during the course of his period in Liverpool, Van Velzen was guilty of an extensive list of activities, including illegally retaining patient tissues, falsifying data and post-mortem reports, and poor record-keeping. Despite this, he was never convicted of any criminal offence; the Human Tissue Act 1961 did not provide any criminal penalties; nor could general criminal laws relating to the offences of misconduct in a public office and dishonesty be brought due to technical difficulties: 'there could be no guarantee that organs which remained in the containers were those originally taken at post mortem examination',[54] and 'there was no prospect of being able to establish that a criminal offence had been committed'.[55] Thus the criminal sanctions of the 2004 Act enacted to deal with such a scenario 'were deemed to be necessary to cater for flagrant and egregious breaches, characterised by the activities of Professor Van Velzen, although intended almost exclusively as a deterrent'.[56]

The only criminal charges that have been brought under the HTA 2004 were in June 2006, when an individual was charged with 'inviting the supply of human material for transplant', having tried to sell his kidney for £24,000 on the Internet.[57] He pleaded guilty, received a twelve-month custodial sentence, suspended for two years, and was ordered to pay £250 in costs. Judge Challinor stated: 'There is much public disquiet surrounding the sale of body parts and an understandable revulsion at such practices. A sentence incorporating an element of deterrence is required.'[58]

The HTA is responsible for making sure that licensed establishments comply with the requirements of the Human Tissue Act 2004, the Human Tissue (Quality and Safety for Human Application) Regulations 2007 and the HTA codes of practice. As part of its statutory remit, the HTA licenses establishments, carries out inspections and conducts

[53] The Royal Liverpool Children's Inquiry, *Report*. [54] Ibid., p. 30, para. 1.10.

[55] 'CPS Advises No Prosecution of Prof. Dick Van Velzen over Alder Hey' (15 December 2004) (see www.cps.gov.uk/news/press_releases/149_04).

[56] Price, 'The Human Tissue Act', 809.

[57] 'Salesman Faces Jail for Trying to Sell Kidney on the Internet', *London Evening Standard*, 22 April 2007 (see www.thisislondon.co.uk/news/article-23393455-salesman-faces-jail-for-trying-to-sell-kidney-on-internet.do).

[58] 'Suspended Jail Term for Gambler Who Tried to Sell Kidney on Net', *The Guardian*, 12 May 2007 (see www.guardian.co.uk/uk/2007/may/12/gambling.ukcrime).

investigations to ensure compliance. Where regulatory breaches are identified, appropriate regulatory action is taken in line with this enforcement policy. It is not clear how many, if any, non-prosecuted breaches of the Human Tissue Act there have been thus far.

As with the law governing the regulation of assisted reproduction, in the context of human biomaterials, Parliament is willing, in theory, to invoke criminal sanctions against some types of activity – yet in practice this has seldom occurred. This could imply that the sanctions are acting as a successful deterrent and the activities prohibited are not occurring, or it could be that, as in the domain of assisted reproduction, breaches occur but are seldom prosecuted. If the latter is the case, this could stem from a tacit acknowledgement that perhaps resort to the 'jackboots of the criminal law'[59] may be too draconian in such cases. When, then, *is* the use of the criminal law appropriate, and what is its role? We turn now to the question of whether the criminal justice process is the appropriate mechanism to govern these areas of science.

Is criminal law appropriate?

> The prohibition inherent in a criminal statute is dual: it addresses the citizen in identifying the proscribed behaviour, and it addresses the organs of the state in establishing conditions precedent to punitive reaction.[60]

We have seen that scientists' responsibilities are often a manifestation of more general moral responsibilities as they apply to those engaged in the business of science; specific regulations have been developed for some areas of particular significance (albeit with varying degrees of success). The question arises, therefore, as to why there is a need for additional regulation of scientists' activities outside the legal framework that governs the conduct of ordinary citizens, or for regulation of medical research beyond the laws that apply to health care professionals in other areas. Of most interest for our purposes, is there anything that specifically merits the invocation of the criminal law? As we have seen, regulation of science overall does not rely heavily on criminal law, and indeed where specific criminal provisions are present, they are seldom invoked. What, then, triggers the need for extra regulation in particular areas of research?

It is clear that human reproduction and use of human biomaterials are emotive subjects, and the use of criminal sanctions in the statutes

[59] A phrase borrowed from Ronald Dworkin, *Life's Dominion: An Argument about Abortion and Euthanasia* (London: HarperCollins, 1993), Chapter 15.

[60] D.A. Thomas, 'Form and Function in Criminal Law' in P.R Glazebrook (ed.), *Reshaping the Criminal Law: Essays in Honour of Glanville Williams* (London: Stevens & Sons, 1978), p. 21.

regulating both came amidst great controversy. The HFE Act was born amidst concerns regarding the morality of research on the human embryo and the ability to manipulate human creation; the Human Tissue Act stemmed from the unlawful removal and retention of human tissue and organs without the appropriate consent. But is criminal law the appropriate mechanism to govern these contentious areas?

Criminal law as moral dictum

> Systems of substantive criminal law serve several purposes. Politically a criminal code may acquire symbolic significance as an expression of national unity ... Morally, the code may amount to a concrete manifestation of the judgement of the community on the central values which bind it together and serve notice on the citizen of the limits of permissible behaviour within that society.[61]

One possible justification for the use of criminal law to regulate science is that criminal law is often used as an instrument to dictate what society deems to be morally acceptable. Research on human reproduction and biomaterials has long attracted controversy and debate, and there may never be a clear consensus on the complex ethical and social issues involved.

The criminal law might thus provide a legislative mechanism for dealing with controversial research techniques and offering guidance as to what society deems to be acceptable or legitimate behaviour. Given the ongoing moral controversy and widespread disagreement over what is acceptable, however, is it appropriate to brand activity on one side with the stigma of criminalisation?

Hindering scientific progress

An objection to the use of criminal law to regulate scientific research is that it could stifle innovative research and deprive us of the benefits of medical and scientific progress. The fact that the UK has been stated to have 'an unrivalled position in the world for scientific research and innovation' suggests that the regulatory framework has struck an appropriate balance and the criminal sanctions are not unduly stifling science;[62] stem cell science in the UK is flourishing.[63]

[61] Ibid., p. 21.

[62] Better Regulation Task Force, *Scientific Research: Innovation with Controls* (London: Cabinet Office, 2003).

[63] The world's first embryonic stem cell bank was opened in Hertfordshire on 19 May 2004 and therapeutic cloning was authorised later that year; more recently, licenses have been granted to carry out hybrid embryo research (see www.hfea.gov.uk/519.html).

Whether science has been hindered or not is difficult to ascertain, however, for as Furness notes, 'research that does not occur remains invisible, and nobody will know what we have lost'.[64] He illustrates this point by reminding us of the 2005 Nobel Prize in Medicine, awarded for the discovery (based on human tissue studies) of *Helicobacter pylori* as the main cause of peptic ulcers, and questions whether this research would be possible today:

> Would our new system for tissue regulation still pass this *Helicobacter* test? The letter of the law passes the test, but the system may still fail as a result of researchers being deterred by fear or bureaucracy, or by ethics committees demanding unobtainable consent. Sadly we will never know if the next *H pylori* discovery is blocked.[65]

That criminal sanctions would stifle innovative research was certainly a real fear when the Human Tissue Bill was going through Parliament. The Human Tissue Act 2004 as initially drafted 'drew a vociferous response from the pharmaceutical, pathology, medical research and research charity communities in particular, alleging that vital research would be compromised'.[66]

Discouraging research

Clearly the threat of the criminal law invokes fear amongst researchers, though the effects of this are hard to prove. A related issue, then, is whether the threat of criminal prosecution might pose a disproportionate obstacle or deterrent to legitimate research. In relation to the Human Tissue Act 2004, Furness notes that the legislation is complex, and scientists dread criminal sanctions. 'This induces a tendency to "gold plate" the legislation; the response becomes "if in doubt, don't do it". I am repeatedly told by colleagues that the Act prohibits things that in fact it doesn't, or demands things that it doesn't, or that a licence is required when it isn't.'[67] Uncertainty over the law and trepidation over possible criminal prosecution may discourage potentially beneficial research.

[64] P. Furness, 'The Human Tissue Act: Reassurance for Relatives, at a Price' (2006) 333 *British Medical Journal*, 512.
[65] Ibid.
[66] 'Human Tissue Bill Threatens Medical Research, Say Charities', *The Times*, 16 February 2004; 'Human Tissue Bill Will Outlaw Research Warns Drug Firm', *The Guardian*, 24 February 2004; 'Risk to Research in Human Tissue Bill', *The Times*, 21 February 2004.
[67] Furness, 'The Human Tissue Act', p. 512.

Legal regulation and scientific freedom

The potential of criminal regulation to block research highlights an underlying tension between the (perceived) need for regulation and the importance of scientific freedom. The freedom to pursue research is seen as an essential element of science, necessary to realise the benefits of creative inquiry and to maintain its independence and impartiality, while regulation often limits that freedom in certain ways.

It must be noted, however, that the principle of scientific freedom means not that scientists should have carte blanche to do whatever they like in the name of research, but that certain liberties are necessary for science to progress and should therefore be protected, such as 'freedom of movement, association, expression and communication for scientists as well as equitable access to data, information and research materials'.[68] Restrictions on these may have a detrimental, sometimes unjust effect on the direction of science and on those engaged in research.[69] Not all regulation, though, is necessarily antithetical to scientific freedom.

The question remains whether it is appropriate for the law to express, in effect, a position that some sorts of research should not be done at all. The justifications for maintaining scientific freedom suggest that there may be serious problems with allowing the reach of the law to extend into not just how research should be done, but on what. Current regulation in some areas, including reproductive biomedicine, comes very close to doing just that. The ban on therapeutic cloning that exists in several countries (though not the UK) has been described as 'an unprecedented interference by government in scientific and academic inquiry'.[70] Even without an explicit ban, if the only route of inquiry into a research question is via a technique that is declared unlawful, in practice there will be no way to pursue such an investigation. To take one example from the HFE Act, the prohibition against culturing human embryos past fourteen days is likely to make research into ectogenesis (*in vitro* gestation) almost impossible.

There are of course other forms of indirect regulation that can have an impact on scientific freedom – for example, how funding is allocated can have a strong influence on the kind of research that scientists choose to pursue; but we are concerned here with direct criminal regulation of scientific activity, which has a more immediate, if not necessarily greater,

[68] International Council for Science, *Freedom, Responsibility*, p. 5.

[69] For example, restrictions on international movement and communication can disadvantage researchers from particular countries.

[70] T. Caulfield, 'Scientific Freedom and Research Cloning: Can a Ban Be Justified?' (2004) 364 (9429) *Lancet*, 124.

impact. There may well be reasons to impose some limitations on scientific activity, despite the strong presumption of scientific freedom. As Caulfield puts it, however:

> [T]he concept of scientific freedom serves as a lens through which to view the legitimacy of a given regulatory approach. It places the onus of justification on those seeking to curtail scientific inquiry and should force us to consider whether there are other, less intrusive, regulatory instruments that could be used to achieve the policy objectives.[71]

In light of this, how should we justify the use of criminal law, potentially one of the most intrusive and severe forms of regulation, in some areas of science? The key perhaps lies in that potential, as we have seen, rarely being invoked: perhaps the criminal law forms another sort of lens, a cautionary and symbolic one, through which to view scientific activity.

Conclusions: moral controversy and criminal law – a symbolic role?

Negotiating the legitimate bounds of scientific activity requires a precarious balancing act. On one hand there is the desire to encourage innovative research, whilst on the other the need to minimise harms that research might cause, and ensure accountability so as to allay public fears. Regulators must grapple with these opposite considerations:

> [The] law must adequately reflect two competing (and often conflicting) public policy considerations: on the one hand it must facilitate research in view of the gains in knowledge that may be made in the fields of medical/biological science and the improvements this may promise for human health; whilst on the other it must repress such research as is regarded as unacceptable, primarily due to doubts about its ethical justifiability.[72]

Some forms of scientific regulation are aimed at the prevention of harms, feared or foreseen. The legislation regarding genetically modified organisms discussed above is one such example, where the feared harm stems from the possible adverse consequences of uncontrolled release of GMOs into the wild, or unchecked exposure of humans to GMOs. The justification in such cases for state interference in the form of law is the possibility of serious harm to others, a position that accords with the liberal philosophical principles underlying our political and legal system.

[71] Ibid.
[72] R. Brownsword, 'Regulating Human Genetics: New Dilemmas for a New Millennium', as cited by S. Halliday and D. Lyn Steinberg, 'The Regulated Gene: New Legal Dilemmas' (2004) 12(2) *Medical Law Review*, 2.

When it comes to regulation in areas of moral controversy, it is perhaps less clear that research will cause harm. Taking the example of human embryo research, whilst some might regard the destruction of embryos as a morally significant harm, there are some who believe that embryos lack moral status and harms to them are therefore of no moral importance. No moral consensus on this issue has yet been reached, nor is it likely in the near future. The justification for regulation is not to prevent a harm universally acknowledged. What, then, is the role of the state and, in particular, the criminal law in regulating these areas of research?

As the Warnock Report observed, there is a need for some authority to assume 'responsibility to regulate and monitor practice in relation to those sensitive areas which raise fundamental ethical questions'[73] – not to prevent or curtail absolutely research in these fields, but to keep science in touch and in step with the concerns of society and to facilitate the relationship between them. In these areas, therefore, rather than serving to prevent acts that may result in harms, regulation may serve as an expression of the effective 'social contract' between science and society. On this interpretation, it is the element of ethical oversight (of science being responsive to the needs and views of society, which in turn supports and facilitates science) that forms the key part of such regulation. As Brazier comments, regarding the introduction of the HFE Act 1990 to regulate embryo research:

> Unregulated embryo research was simply not an option ... Regulation was the price for ensuring the 'legitimacy' of such research.[74]

This being the case, we might expect that compliance measures would generally be aimed at ensuring participation in this agreement as to how science (or a certain type of science) is governed, rather than preventing research altogether. The stigma of criminalisation attaches not to the controversial research itself, but to its illegitimate, unregulated pursuit. Indeed, with respect to the HFE Act, this is mostly the case: although there is still no moral consensus on the acceptability of destructive embryo research, the Act allows such research under its licence terms, but imposes penalties for conducting such research without a licence – that is, outside the Authority's oversight.

At least in the case of research on human reproduction and biomaterials, then, criminal law is used less as a direct regulatory mechanism than

[73] Department of Health and Social Security, 'The Warnock Report'.
[74] M. Brazier, 'Regulating the Reproduction Business' (1999) 7(2) *Medical Law Review* 166, 172.

to fulfil the requirement of gaining public trust and legitimising science that would, if not subjected to the powerful symbolic force of criminal regulation, arouse even greater public concern. The increasing willingness of prosecutors to criminalise scientific misconduct may play a similar role, attempting to restore public trust in science in the face of an individual scientist's breach of that trust. Perhaps in other areas of research, the science–society relationship can be likewise moderated.

9 Bioethical conflict and developing biotechnologies: is protecting individual and public health from the risks of xenotransplantation a matter for the (criminal) law?

Sara Fovargue

Introduction

Some (bio)medical techniques, procedures and developments demand the support of the law, particularly the criminal law, because of the risks they pose. Xenotransplantation (the transplantation of live non-human animal organs, tissues or cells into humans) is one such biotechnology, and my focus here is on genetically engineered pig solid organ xenotransplants. Such xenotransplants raise bioethical conflict because, if successful, they would benefit individuals and society, but there are risks. By (xeno)transplanting a non-human animal organ into a human, infectious diseases might be transmitted to the recipient and others. It has thus been suggested that the xeno-recipient, her contacts and others should be subject to lifelong surveillance to manage those risks. In this essay I highlight some of the key aspects of the UK's draft surveillance scheme, one of the first published, and emphasise the lack of consideration of sanctions for non-compliance. This is curious because, without such sanctions, any surveillance scheme is meaningless and cannot protect individual and public health.[1] However, if clinical xenotransplants are to proceed, the risks necessitate the involvement of the law to help preserve health, and I explore the different forms of law that might be utilised to achieve this. I suggest that, despite understandable concerns about the dangers of overcriminalisation and threats to established and legally protected human rights,[2] the risks of xenotransplantation are such that the intervention

[1] For discussion of the meaning of 'public health', see Coggon in this volume.

[2] E.g. B. McSherry, A. Norrie and S. Bronitt, *Regulating Deviance: The Redirection of Criminalisation and the Futures of Criminal Law* (Oxford: Hart, 2009); D.N. Husak, *Overcriminalization: The Limits of the Criminal Law* (Oxford University Press, 2008).

of the *criminal* law is imperative. If this is an unpalatable and unacceptable conclusion, then moving xenotransplants from the laboratory to the clinic is risky and dangerous and requires particularly careful justification, as, without these safeguards, public health may be jeopardised for (possible) private benefit.

An overview of the risks

For a xenotransplant to support human life, immunological, physiological and microbiological barriers must be addressed.[3] With the latter, xenotransplanting a pig organ into a human might transfer infectious diseases across the species barrier to the xeno-recipient, and on to her contacts and the public. Pigs can be bred and raised in specific pathogen-free environments and the biological differences between the species may decrease the risk of transferral; however, genetically engineering the pigs may actually increase the risk of transmission. There is limited information on which diseases may be transferred, and the methods for detecting them are limited because no genetically engineered solid organ pig-to-human xenotransplant has yet been performed.[4] Nevertheless, since it was discovered during the 1990s that porcine endogenous retroviruses (PERVs) could infect human cells *in vitro*, and, under certain circumstances, actively infect human cells, this retrovirus has been of most concern.[5] Retroviruses have been linked to conditions involving an impaired immune system,

[3] On the immunological and physiological barriers, see, e.g., B. Petersen, J. W. Carnwath and H. Niemann, 'The Perspectives for Porcine-to-Human Xenografts' (2009) 32 *Comparative Immunology, Microbiology and Infectious Diseases*, 91; J. M. Dobson and J. H. Dark, *The Physiology of Xenotransplantation* (London: Department of Health, 2002), Chapter 6.

[4] On the microbiological risks, see, e.g., J. L. Whitteker, A. K. Dudani and E. S. Tackaberry, 'Human Fibroblasts Are Permissive for Porcine Cytomegalovirus *In Vitro*' (2008) 86 *Transplantation*, 155; N. Griesche, D. Zikos, P. Witkowshi, A. Nitsche, H. Ellerbrok, O. B. Siller, G. Pauli and B. Biere, 'Growth Characteristics of Human Adenoviruses on Porcine Cell Lines' (2008) 373 *Virology*, 400.

[5] V. Specke, S. Rubant and J. Denner, 'Productive Infection of Human Primary Cells and Cell Lines with Porcine Endogenous Retroviruses' (2001) 285 *Virology*, 177; U. Martin, M. E. Winkler, M. Id, H. Radecke, L. Arseniev, R. Grotelüschen, A. R. Simon and G. Steinhoff, 'Transmission of Pig Endogenous Retrovirus to Primary Human Cells' (2000) 32 *Transplantation Proceedings*, 1157; U. Martin, M. E. Winkler, M. Id, H. Radeke, L. Arseniev, Y. Takeuchi, A. R. Simon, C. Patience, A. Haverich and G. Steinhoff, 'Productive Infection of Primary Human Endothelial Cells by Pig Endogenous Retrovirus' (2000) 7 *Xenotransplantation*, 138; P. Le Tissier, J. P. Stoye, Y. Takeuchi, C. Patience and R. A. Weiss, 'Two Sets of Human-Tropic Pig Retrovirus' (1997) 389 *Nature*, 681; C. Patience, Y. Takeuchi and R. A. Weiss, 'Infection of Human Cells by Endogenous Retrovirus of Pigs' (1997) 3 *Nature Medicine*, 282.

neurodegeneration and cancers of the blood system;[6] and the viruses that cause AIDS in humans resulted from retroviral cross-species transmissions from non-human primates.[7] Thus, '[e]vidence from naturally occurring retroviral zoonosis and cross-species infections by animal retroviruses, for example, HIV, provides a basis for reasoned speculation on the risk posed by PERVs. In a worst case scenario xenograft-related PERV transmission would be the starting point of a new viral disease resulting in a public health problem.'[8] What is more, the greatest risk may come from 'non-PERV unknown viruses that may be non-pathogenic in source animals but could adapt and become pathogenic in immunosuppressed humans'.[9] So, whilst pigs may be bred free from specific known organisms, 'newly emerging or unrecognised organisms, as well as those that cannot be eliminated (currently endogenous retroviruses for example) continue to pose a potential threat'.[10]

Without clinical xenotransplants it is hard to determine the effect of PERVs and other diseases, their consequences and the magnitude of risk, but experience with other retroviruses suggests that more than flu-like symptoms are likely and clinical xenotransplants could introduce a lethal infectious disease into society. The possibility of such transmission has been acknowledged by some xenotransplant regulators, which have proposed surveillance regimes for xeno-recipients, their close contacts and relevant health professionals.[11] The very existence of such regimes suggests that although the nature and extent of xenotransplant risks are uncertain, they are not to be taken lightly.

[6] V. Specke, R. Plesker, J. Wood, C. Coulibaly, K. Suling, C. Patience, R. Kurth, H.-J. Schuurmann and J. Denner, 'No In Vivo Infection of Triple Immunosuppressed Non-Human Primates after Inoculation with High Titers of Porcine Endogenous Retrovirus' (2009) 16 *Xenotransplantation*, 34; B. Dieckhoff, B. Petersen, W. A. Kues, R. Kurth, H. Niemann and J. Denner, 'Knockdown of Porcine Endogenous Retroviruses (PERV) Expression by PERV-Specific shRNA in Transgenic Pigs' (2008) 15 *Xenotransplantation*, 36.

[7] Dieckhoff *et al.*, 'Knockdown', 37.

[8] D. Louz, H. E. Bergmans, B. P. Loos and R. C. Hoeben, 'Reappraisal of Biosafety Risks Posed by PERVs in Xenotransplantation' (2008) 18 *Reviews in Medical Virology*, 53, 60.

[9] M. Sykes, '2007 IXA Presidential Address: Progress Toward an Ideal Source Animal: Opportunities and Challenges in a Changing World' (2008) 15 *Xenotransplantation*, 7, 10.

[10] D. A. Muir and G. Griffin, *Infection Risks in Xenotransplantation* (London: Department of Health, 2001), p. 118.

[11] Including Health Canada, *Proposed Canadian Standard for Xenotransplantation* (Ottawa: Health Canada, 1999); United States Department of Health and Human Services, Food and Drug Administration, *PHS Guideline on Infectious Disease Issues in Xenotransplantation* (Rockville, MD: FDA, 2001) (see www.fda.gov/downloads/BiologicsBloodVaccines/GuidanceComplianceRegulatoryInformation/Guidances/Xenotransplantation/UCM092858.pdf).

Xeno-surveillance

In 1999, the United Kingdom Xenotransplantation Interim Regulatory Authority (UKXIRA), the then xenotransplant regulator, published draft guidance on xeno-surveillance.[12] Although the UKXIRA was disbanded in 2006, this draft remains valid guidance,[13] and under it information should be collected on and from the non-human animals, those working with them, xeno-recipients, their 'household/close contacts'[14] and relevant health professionals.[15] A xeno-recipient is to consent to the xenotransplant *and* to the 'post-operative procedures and constraints' with which they have 'a *commitment* to comply'.[16] Weekly, monthly and yearly schedules for minimum pre- and post-xenotransplant *lifelong* recipient screening are set out,[17] and data and samples will be stored 'indefinitely'.[18] It is '*likely*' that xeno-recipients will be '*require[d]*' to comply with the surveillance scheme, regardless of the operation's success, and remain contactable post-xenotransplant.[19] They '*must* refrain' from donating blood or organs, '*must*' agree to a post mortem, tell family members of this and agree to post-mortem samples being stored, with no time set.[20] They '*should*' seek advice before having children and barrier contraception is '*recommended*'.[21] The recipient's household/close contacts will be seen by the xeno-team prior to the operation, baseline blood samples taken, and samples and records maintained and archived, possibly indefinitely.[22] They '*should*' be subject to the same limits on donation as xeno-recipients, '*should*' know their GPs will be aware of their status, and '*may*' need to remain contactable.[23] Active clinical surveillance will only occur via the recipient's scheduled follow-up visits and the reporting of any 'untoward events'; they are not themselves independently clinically monitored.[24] For relevant health

[12] UKXIRA, *Draft Report of the Infection Surveillance Steering Group of the UKXIRA* (London: Department of Health, 1999).
[13] Personal communication with the Department of Health (9 August 2007) (on file with author).
[14] '[H]ousehold members and sexual partners and others with whom the xenotransplant recipient may engage in activities in which bodily fluid may be exchanged' (UKXIRA, *Draft Report*, para. 2.5 and Annex Seven).
[15] Ibid., para. 2.5. [16] Ibid., A5.4 (emphasis added).
[17] Ibid., para. 2.5 and Annex Two. [18] Ibid., A5.2.
[19] Ibid. (emphasis added). [20] Ibid. (emphasis added).
[21] UKXIRA, *Third Annual Report September 1999–November 2000* (London: Department of Health, 2001), para. 5.9 (emphasis added).
[22] UKXIRA, *Draft Report*, para. 2.5 and A5.2. [23] Ibid., A5.2 (emphasis added).
[24] Ibid., para. 2.8 and Annex Three. 'Untoward events' are 'suspected and confirmed xenogeneic infections occurring in recipients and their close contacts and in healthcare and animal workers' (ibid., Annex Seven).

professionals,[25] pre-xenotransplant baseline blood samples '*should*' be taken, archived indefinitely[26] and subject to laboratory tests if there is a 'significant exposure'[27] to a xeno-recipient's bodily fluids or an untoward event.

Compliance

Receiving or performing a xenotransplant will clearly have lifelong implications, and although it is usual for any organ recipient to be monitored for life, with a xenotransplant the net of surveillance may be wider. The extensive nature of the draft scheme indicates not only the seriousness with which the risks are taken, but also that complying with its requirements is 'presumably of the highest order of significance otherwise it would not be required in the first place'.[28] Consent to the xenotransplant and to the 'post-operative procedures and constraints' 'will be needed *explicitly* and *in advance*' from xeno-recipients and, as relevant, their contacts,[29] but the consent of health professionals was not addressed. The UKXIRA highlighted the importance of assessing the likely compliance of potential recipients in the selection process,[30] along with their 'commitment to comply' with the scheme,[31] and stated that only *voluntary* compliance with surveillance was possible because no UK legislation required or forced patients to do so.[32] Notably, legislation to that effect was 'probably not needed' because xeno-recipients '*will normally* perceive the need to attend regular check-ups and to comply with an immunosuppressive drug regime to maintain their health'.[33] If, however, 'a person had developed a highly infectious disease, not previously seen in humans, and which could be named or described with certainty', public health legislation could be amended to allow for medical examinations to be

25 'The xenotransplant team and others at risk of significant exposure to a recipient's body fluids' (ibid., para. 2.5).

26 Ibid., para. 2.5 and Annex Three (emphasis added).

27 Ibid., para. 2.5. An 'exposure to xenotransplant tissue or bodily fluids of recipient. The nature of the exposure needs to be defined further but at least includes percutaneous exposure' (ibid., Annex Seven).

28 S. A. M. McLean and L. Williamson, 'Xenotransplantation: A Pig in a Poke?' (2004) 57 *Current Legal Problems*, 443, 459.

29 UKXIRA, *Draft Report*, A5.4 (original emphasis).

30 Ibid., A5.7. 31 Ibid., A5.4. 32 Ibid., A5.3.

33 Ibid. (emphasis added). Note that this is not in evidence for all transplant recipients, and in other health contexts see, e.g., L. B. Hilbrands, A. J. Hoitsma and R. A. P. Koene, 'Medication Compliance after Renal Transplantation' (1995) 60 *Transplantation*, 914; J. Dunbar-Jacob and M. K. Mortimer-Stephens, 'Treatment Adherence in Chronic Disease' (2001) 54 *Journal of Clinical Epidemiology*, S57.

ordered for patients and carriers,[34] along with their compulsory removal to and detention in hospital.[35] But this legislation would only be operative if the disease were identifiable, present and highly infectious, and was not intended 'to extend to surveillance of those who *might* possibly develop infectious diseases'.[36]

Securing compliance

The possibility of a xeno-recipient and/or others agreeing to comply with surveillance pre-xenotransplant and not subsequently doing so was not substantially addressed in the UKXIRA's 1999 *Draft Report*, with the discussion primarily limited to the applicability of existing public health laws.[37] Legal commentators in the USA and Canada have considered how compliance could be ensured with their surveillance schemes,[38] but there is a sense within some of the scientific literature that these issues should *only* be addressed once the science has been mastered.[39] There is also an assumption that effective, legal, ethical and practical surveillance *can* be established,[40] but this is questionable, as 'there are limits to the enforcement measures that the law will be prepared to accommodate'.[41] Nevertheless, the International Xenotransplantation Association has suggested that 'governments may consider whether research subjects should be informed that certain government services will be withheld from subjects who do not honor their original promises to abide by practices that will minimize

[34] Referring to the Public Health (Control of Disease) Act 1984.
[35] UKXIRA, *Draft Report*, A5.3. [36] Ibid. (emphasis added).
[37] UKXIRA, *Draft Report*, A5.4.
[38] E.g. J. A. Holland, 'The "Catch-22" of Xenotransplantation: Compelling Compliance with Long-Term Surveillance' (2007) 7 *Houston Journal of Health Law and Policy*, 151; P. S. Florencio and E. D. Ramanathan, 'Legal Enforcement of Xenotransplantation Public Health Safeguards' (2004) 32 *Journal of Law, Medicine and Ethics*, 117; P. S. Florencio and E. D. Ramanathan, 'Are Xenotransplantation Safeguards Legally Viable?' (2001) 16 *Berkeley Technology Law Journal*, 937; T. A. Caulfield and G. B. Robertson, 'Xenotransplantation: Consent, Public Health and Charter Issues' (2001) 5 *Medical Law International*, 81.
[39] E.g. H. Y. Vanderpool, 'The International Xenotransplantation Association Consensus Statement on Conditions for Undertaking Clinical Trials of Porcine Islet Products in Type 1 Diabetes – Chapter 7: Informed Consent and Xenotransplantation Clinical Trials' (2009) 16 *Xenotransplantation*, 255, 261.
[40] E.g. E. Cozzi, M. Tallacchini, E. B. Flanaghan, R. N. Pierson, M. Sykes and H. Y. Vanderpool, 'The International Xenotransplantation Association Consensus Statement on Conditions for Undertaking Clinical Trials of Porcine Islet Products in Type 1 Diabetes – Chapter 1: Key Ethical Requirements and Progress Toward the Definition of an International Regulatory Framework' (2009) 16 *Xenotransplantation*, 203.
[41] Florencio and Ramanathan, 'Legal Enforcement', 118.

infectious disease risks'.[42] Whilst the ethics and legality of this is questionable, 'lesser' penalties, such as fines for non-compliance or financial incentives for compliance, could be introduced.[43] The effectiveness of these is debatable, but if public health is to be protected, then more than voluntary compliance and reliance on the goodwill or vagaries of human nature is required. The risks of clinical xenotransplants are such that the law *must* have a role to protect public health, yet this is contentious because, as with other communicable infectious diseases, the xeno-recipient can be seen as an individual 'victim', with needs, interests and rights, *and* as a potential disease vector – a threat to the wider community.[44] It also runs counter to the unspoken idea that the law should have a limited role in health care matters. It is thus important to consider carefully which branch of the law can most effectively ensure compliance with the necessary xeno-surveillance regimes, particularly because here the law needs to be able to secure compliance *prior* to confirmed infection, as ensuring compliance *after* an infection is established may well be too late.

Civil law – contract

In order to safeguard public health, 'something more binding and contractual' than traditionally conceived consent could be introduced; something 'which then would dramatically alter the traditional doctor–patient relationship'.[45] Xeno-recipients and involved others could be required to consent to signing a contract agreeing to comply with the post-xenotransplant surveillance regime.[46] For this to be legally binding, there must be an agreement expressed in a sufficiently certain form

[42] M. Sykes, A. d'Apice and M. Sandrin, 'Position Paper of the Ethics Committee of the International Xenotransplantation Association' (2003) 10 *Xenotransplantation*, 194, 198.

[43] Florencio and Ramanathan, 'Legal Enforcement', 119.

[44] C.B. Smith, M.P. Battin, J.A. Jacobsen, L.P. Francis, J.R. Botkin, E.P. Aslund, G.J. Domek and B. Hawkins, 'Are there Characteristics of Infectious Diseases that Raise Special Ethical Issues?' (2004) 4 *Developing World Bioethics*, 1, 2.

[45] A.S. Daar, 'Ethics of Xenotransplantation: Animal Issues, Consent, and Likely Transformation of Transplant Ethics' (1997) 21 *World Journal of Surgery*, 975, 977.

[46] Suggested, e.g., by Holland, 'The "Catch-22"', 173–4; M.A. Clark, 'This Little Piggy Went to Market: The Xenotransplantation and Xenozoonose Debate' (1999) 27 *Journal of Law, Medicine and Ethics*, 137; A.S. Daar, 'Xenotransplantation: Science, Risk and International Regulatory Efforts' in T.A. Caulfield and B. Williams-Jones (eds.), *The Commercialization of Genetic Research: Ethical, Legal and Policy Issues* (New York: Kluwer Academic, 1999), p. 129; F.H. Bach, J.A. Fishman, N. Daniels, J. Proimos, B. Anderson, C.B. Carpenter, L. Forrow, S.C. Robson and H.V. Fineberg, 'Uncertainty in Xenotransplantation: Individual Benefit versus Collective Risk' (1998) 4 *Nature Medicine*, 141.

for the courts to be able to enforce it and supported by consideration.[47] There may also be some requirements as to the form that the agreement should take, and in England and Wales the parties must have an intention to create legal relations.[48] Both parties must have the capacity to enter into a legal contract, must consent to do so freely and voluntarily, and the purpose of the contract must not be illegal or contrary to public policy.[49]

On the face of it, a contract to abide by the necessary surveillance regime for the xeno-recipient, her contacts and involved health professionals could be drawn up. Some issues would need clarification, such as between whom the contract existed – the hospital in which the operation was performed, the regulator, the state or the company that supplied the non-human source animals. However, a binding contract to ensure participation with lifelong surveillance would be 'inconsistent with current legal principles', because consent, to medical treatment at least, is traditionally characterised as a permission and not a contract.[50] Introducing such a contract would necessarily change the form and nature of the doctor–patient relationship under the National Health Service in the UK. Furthermore, xeno-recipients, for example, could argue post-xenotransplant that they did not have the capacity to consent to a surveillance contract or that they had been coerced into signing it because of the situation they were in; with no other hope.[51] There may also be policy reasons for not viewing consent as a binding contract, including this vulnerability and the balance of knowledge and power between doctor and patient; and such a contract may be legally unenforceable as being against public policy because of the necessary restrictions on liberty it must contain.[52] Furthermore, if the xeno-recipient and/or others could show that they were unduly influenced to enter into the contract because the other party to the contract took advantage of the relationship of

[47] With regard to England and Wales, see, e.g., E. McKendrick, *Contract Law: Text, Cases and Materials* (4th edn) (Oxford University Press, 2010), pp. 19–20.

[48] Ibid.

[49] E. McKendrick, *Contract Law* (8th edn) (Basingstoke: Palgrave Macmillan, 2009), Chapters 15–16.

[50] Caulfield and Robertson, 'Xenotransplantation', 86.

[51] It has been suggested that the first xeno-recipients should be in this position; see, e.g., S. Welin and M. S. Sandrin, 'Some Ethical Problems in Xenotransplantation: Introductory Remarks at Ethics Workshop' (2006) 13 *Xenotransplantation* 500; Z. Ibrahim, M. Ezzelarab, R. Kormos and D. K. C. Cooper, 'Which Patients First? Planning the First Clinical Trial of Xenotransplantation: A Case for Cardiac Bridging' (2005) 12 *Xenotransplantation*, 168.

[52] *Horwood* v. *Millar's Timber* [1917] 1 KB 305.

trust and confidence in her, to the substantial detriment of the xeno-recipient, and so on, the contract could be set aside.[53]

Even if these issues are addressed, the effectiveness of the civil law to protect individual and public health will rest on its ability to enforce *compliance* with the surveillance contract. If the xeno-recipient, for example, breaches that contract by refusing to comply with the surveillance regime, monetary damages could be claimed for loss caused by the breach; with the aim being to put the claimant in the position she would have been in had the contract been performed.[54] But the claimant can only be the other party to the contract, and it may not be this party who suffers loss by being infected because of the breach. Even if this is the infected and affected party, the damage has already been done and, as with all contractual breaches that result in personal injury, no amount of money will put the claimant in the position she would have been in if the breach had not occurred. Additionally, the damage from the xeno-transmitted infection may not be evident for some time after the breach, so the xeno-recipient's lifelong liability would need clarifying, and it might prove difficult to establish causation after such a lapse of time. It is also unclear whether a court would utilise the equitable remedy of specific performance and use its discretion to order the person breaching the surveillance contract to comply with it, because the role of specific performance as a remedy is, in England and Wales at least, ambiguous.[55] Regardless, specific performance after the fact may protect the health of others, but not those who have already been infected because of the breach. In the light of these concerns, the civil law (via a contract) is not an effective way of enforcing compliance with a xeno-surveillance regime, as a contract would be of little value if the signatory could not be *compelled* to follow the regime *prior* to infection being established.

Criminal law

Existing public health provisions Public health laws are clearly designed to protect individual and public health, and in England and Wales such provisions are, as noted below, primarily supported by the criminal law penalty of a substantial fine for non-compliance. With regard to xenotransplantation, the Council of Europe has recommended that non-compliance with surveillance regimes could result in public authorities taking necessary and proportionate 'appropriate measures'

[53] McKendrick, *Contract Law: Text*, Chapter 19.
[54] Ibid., Chapter 23. [55] Ibid., Chapter 24.

to protect public health.[56] The Organisation for Economic Co-operation and Development, however, has suggested that 'it is unlikely that public health officials will be able to enforce strict compliance or restrain young patients in particular who have returned to good health and who are looking forward to leading a normal life',[57] and problems with using US and Canadian public health statutes for asymptomatic individuals have been identified.[58] In some jurisdictions a state of public health emergency would need to be declared for public health statutes to apply,[59] and a key issue is the need for compulsory compliance with post-xenotransplant surveillance to pre-empt, anticipate and prevent transmission; but it is questionable whether public health laws will apply without *proven* infection.[60]

The relevant legislation in England and Wales is the Public Health (Control of Disease) Act 1984, which has been significantly amended since the UKXIRA's 1999 report. Under the amended 1984 Act, the appropriate Minister can make regulations to protect against, prevent, control or provide a public health response to the incidence or spread of infection in England and Wales.[61] This power can be used in relation to infection generally and particular forms of it, and provisions of a general, contingent or specific nature can be made.[62] Health protection regulations may include provisions (a) imposing notification and recording duties on registered medical practitioners for confirmed and suspected infection cases; (b) giving local authorities public health risks monitoring functions; and (c) imposing restrictions or requirements on persons 'in the event of, or in response to, a threat to public health'.[63] The latter include prohibiting or restricting holding an event, and 'a special restriction or requirement' that can only be imposed by a magistrate.[64] For threats to public health, a restriction or requirement under section 45C (3)(c) cannot be provided for or imposed unless the appropriate minister, and then the decision-maker, considers that those measures are 'proportionate to what is sought to be achieved by imposing it'.[65]

56 Article 21 Council of Europe, Recommendation Rec(2003)10 of the Committee of Ministers to member states on xenotransplantation.
57 Organisation for Economic Co-operation and Development, *Xenotransplantation: International Policy Issues* (Paris: OECD, 1999), p. 72.
58 E.g. Holland, 'The "Catch-22"', 167–72; Florencio and Ramanathan, 'Legal Enforcement', 119; Caulfield and Robertson, 'Xenotransplantation', 81, 92–93; Florencio and Ramanathan, 'Are Xenotransplantation Safeguards?', 953–60.
59 Florencio and Ramanathan, 'Legal Enforcement', 119.
60 UKXIRA, *Draft Report*, A5.3.
61 Section 45C(1) Public Health (Control of Disease) Act 1984 as amended by Health and Social Care Act 2008.
62 Section 45C(2).
63 Section 43C(3).
64 Sections 45C(4) and (6).
65 Section 45D(1)–(2).

Provisions enabling imposing a special restriction or requirement can *only* be included if there is (i) a 'serious and imminent threat to public health' when the regulations are made, or (ii) the decision to impose the restrictions or requirements in the regulations is 'contingent on there being such a threat at the time when it is imposed'.[66] Regulations can be made for 'preventing danger to public health' because of international travel and from persons arriving at any place,[67] and might include provisions for 'medical examination, detention, isolation or quarantine of persons', 'prohibiting or regulating ... the entry or exit of persons', and 'requiring persons to provide information or answer questions (including information or questions relating to their health)'.[68] Regulations made under sections 45B or C cannot order medical treatment,[69] and although infection refers to 'infection or contamination which presents or could present significant harm to health',[70] an identifiable infection is still required in order for health protection regulations to be made. Importantly, regulations made under sections 45B or C can *only* create an offence punishable by a fine of less than £20,000.[71]

Magistrates have additional powers under the amended 1984 Act and can order, inter alia, that a person submits to medical examination, is removed to or detained in a hospital or other suitable establishment, is kept in quarantine or isolation, provides information, has their health monitored, or is restricted as to where they go or with whom they have contact – a Part 2A Order.[72] Such an order can be made if the magistrate is 'satisfied' that (a) the person 'is or may be infected', (b) with an infection 'which presents or could present significant harm to human health', (c) there is 'a risk' that the person 'might' infect others, and (d) 'it is necessary to make the order in order to remove or reduce that risk'.[73] Magistrates can also order a person who is or may be infected to provide information or answer questions about their health 'or other circumstances (including, in particular, information or questions about the identity of a related party[74])'.[75] This order may be made if (a) a person is or may be infected, (b) the infection is or could present 'significant harm to human health', (c) there is 'a risk' that a related

66 Section 45D(4). 67 Section 45B.
68 Section 45B(2)(b), (e) and (g). Medical examination includes 'microbiological, radiological and toxicological tests' (section 45T(3)).
69 Section 45E. 70 Section 45A(3). 71 Section 45F(5).
72 Section 45G(2). 73 Section 45G(1).
74 Someone who has or may have infected the person, or someone whom the person has or may have infected (section 45G(5)).
75 Section 45G(4).

party 'might' infect others, and (d) it is necessary to make the order to remove or reduce the risk.[76] The Health Protection Regulations 2010 set out what evidence must be available to the magistrate before they can be satisfied that there are grounds for making a Part 2A Order, including a report on, inter alia, the person's diagnosis, test results and summary of the characteristics and effects of the infection.[77] In addition to the restrictions or requirements in the provision under which it is made, the order may also include 'such other restrictions or requirements as the justice considers necessary for the purpose of reducing or removing the risk in question'.[78] Any restriction or requirement in a Part 2A Order *must* have a specified time limit, but this can be extended by a magistrate's order,[79] and if the Part 2A Order imposes detention in a hospital, quarantine or isolation, then the initial period specified in the order or any extension *cannot* exceed twenty-eight days.[80] The appropriate minister can specify in regulations the maximum period for which any restriction or requirement, other than one for detention, in those circumstances may be imposed, and the maximum period of any extension of that.[81] It is an offence, again only punishable with a fine of up to £20,000, to fail to comply without reasonable excuse with a restriction or requirement imposed by or under a Part 2A Order or to wilfully obstruct anyone executing the order.[82] Furthermore, if a person leaves a place at which a Part 2A Order requires them to be detained, isolated or quarantined, the police may take them into custody and return them to that place.[83]

Whilst the amended 1984 Act considers threats to public health, the Act will not assist with xeno-surveillance, as an identifiable infection is required and this may not be possible with xenotransplants. Furthermore, any health protection regulation made under the Act or a Part 2A Order will still not ensure *compliance* with a xeno-surveillance regime, and some of the most relevant provisions are time-limited. The value of imposing a fine for non-compliance or returning a person to their place of detention, isolation or quarantine (for no more than twenty-eight days) is also questionable in this context; thus, utilising the criminal law via public health statutes may prove as ineffective as the civil law in protecting individual and public health post-clinical xenotransplants.

[76] Section 45G(3).
[77] Regulation 4 Health Protection (Part 2A Orders) 2010, SI 2010/658.
[78] Section 45K(2) Public Health (Control of Disease) Act 1984 as amended.
[79] Sections 45K(6), 45L(1)–(2). [80] Sections 45L(3)–(4).
[81] Sections 45L(3)–(4). [82] Sections 45O(1)–(2). [83] Sections 45O(4)–(5).

Existing criminal law provisions The criminal law might, however, have a role to play in protecting health in the xeno-context because of its (debatable) deterrent effect,[84] ability to affect and influence behaviour generally, and symbolic value. Widespread publication of prosecutions and/or punishments may persuade others to comply with xeno-surveillance regimes, but of more practical use may be the criminal law's ability to punish those who do not comply with xeno-surveillance and thereby harm others. However, it has long been accepted that it is difficult to bring a charge of murder or attempted murder under the English criminal law against someone who has infected another with human immunodeficiency virus (HIV),[85] and the same will hold true for xeno-recipients. But the Offences Against the Person Act (OAPA) 1861 might be relevant. Whilst there is no specific offence of transmitting a communicable infectious disease, sections 23 and 24 relate to maliciously administering 'any poison or other destructive or noxious thing' to endanger the life of another or inflict grievous bodily harm on them, or intending to injure, aggrieve or annoy them. Alternatively, an offence under section 47, assault occasioning actual bodily harm,[86] might apply. Here it has to be shown that there was an assault or battery, the victim suffered actual bodily harm, the harm was caused by the defendant's assault or battery, and the defendant intended or was reckless as to whether the victim would suffer an assault or battery.[87] Another possibility is charging the xeno-recipient with an offence under section 20 – unlawfully and maliciously wounding or inflicting any grievous bodily harm on any other person. With no legal justification, the xeno-recipient must have wounded the victim or caused them grievous bodily harm and intended or foreseen that the victim might suffer some harm, not necessarily grievous bodily harm. Successful prosecutions have been obtained under section 20 with regard to the transmission of HIV;[88] so where A knows he is HIV positive, or has some other serious sexual disease, and has consensual sexual intercourse with B, causing B to become HIV positive, A will be guilty of a section 20 offence if he is aware of the risk that by having unprotected sex he may cause B to suffer some harm and B does not

[84] E.g., P. H. Robinson and J. M. Darley, 'Does Criminal Law Deter? A Behavioural Science Investigation' (2004) 24 *Oxford Journal of Legal Studies*, 173.

[85] E.g., D. C. Ormerod and M. J. Gunn, 'Criminal Liability for the Transmission of HIV' [1996] 1 *Web Journal of Current Legal Issues*; S. Bronitt, 'Spreading Disease and the Criminal Law' [1994] *Criminal Law Review*, 21; K. J. M. Smith, 'Sexual Etiquette, Public Interest and the Criminal Law' (1991) 42 *Northern Ireland Legal Quarterly*, 309.

[86] '[A]ny hurt or injury calculated to interfere with the health or comfort' of the victim (*R* v. *Donovan* [1934] 2 KB 498, 509, CCA, *per* Swift J).

[87] *R* v. *Savage*; *R* v. *Parmenter* [1992] 1 AC 699, HL.

[88] E.g. *R* v. *Dica* [2004] EWCA Crim 1103; *R* v. *Konzani* [2005] EWCA Crim. 706.

consent to that risk.[89] Finally, causing grievous bodily harm[90] with intent under section 18 could be relevant. The Court of Appeal has confirmed that if X is HIV positive and has consensual sexual intercourse with Y and this causes Y to become HIV positive, X is guilty of a section 18 offence if X intends to cause Y to suffer grievous bodily harm as a result of the intercourse. It is irrelevant whether Y consents to the risk of HIV infection.[91]

A xeno-recipient who does not comply with the procreation requirements of a xeno-surveillance regime, for example, and thereby infects another, may fall within one of the OAPA provisions. There are, however, problems with applying these sections in this context. For sections 23 or 24 to apply, an infectious disease must be construed as a poison or destructive or noxious thing, and it must be established that the poison or destructive or noxious thing had been 'administered' by the xeno-recipient to a named other.[92] A section 20 offence may be easier to prove if a xeno-recipient is told she may have a communicable infectious disease and not to have unprotected sex, but nonetheless does so without telling her partner that she may be infected. She could argue that she believed there was no risk that she would infect her partner, or that she told her partner of the risk and they had agreed to take that risk; however, as the xeno-recipient will be told from the outset of the risks of unprotected sex, only the latter defence appears relevant here.[93] In contrast, proving intention may not be straightforward, and it is only intentional transmission of a xeno-infection that may constitute a criminal offence under section 18. More generally, the 1861 Act will only apply if an infection is identified in the xeno-recipient and 'victim', and this may not happen for some time because of difficulties in diagnosis, limits of existing tests and problems of latency. It is also debatable whether the OAPA offences only apply where infections are transmitted between sexual partners or in other situations. Furthermore, the causal link between the xeno-recipient and the person infected must be established and this may not be straightforward. Similarly, it is unclear whether the 1861 Act applies where the xeno-recipient has infected another but did not know that they were infected themselves at the relevant time. Are they liable on the basis that they were informed at the time of the xenotransplant of the *risk* of infection, and hence the need

[89] Discussed, for example, in Gurnham in this volume.
[90] '[R]eally serious' harm (*DPP* v. *Smith* [1961] AC 290, 344 HL, *per* Viscount Kilmuir VC).
[91] *Savage*; *Parmenter*.
[92] See, further, Ormerod and Gunn, 'Criminal Liability'.
[93] *Savage*; *Parmenter*.

to comply with the surveillance scheme? Until a case is brought before a court, it is unclear whether any OAPA offences would apply to an infectious disease transmitted post-xenotransplant.

Whilst the criminal law may thus be used *after* the fact to punish the xeno-recipient who infects another, it cannot help to protect public health *in advance*,[94] and using the criminal law in this way could prove counterproductive for public health initiatives aimed at controlling the spread of infectious disease, by stopping people going for testing.[95] Additionally, trying a xeno-recipient for transmitting PERVs to their partner, for example, does little for *public* health; if the infection has already been transmitted then it is too late. To protect public health, action must be taken *before* this point. The existing criminal law's ability to protect public health in this context is thus questionable, and its appropriate use here debatable.

Conclusion: a new xenotransplantation statute

In order to secure compliance with an effective xeno-surveillance regime, new specific xeno-legislation is required. Introducing such an Act would be unusual, as it is not common, in England and Wales at least, for statutes to deal with specific medical procedures or treatments;[96] however, a specific xeno-statute has been introduced or proposed elsewhere.[97] A xenotransplant Act should enforce compliance with the necessary surveillance regime without the need, for example, for declaring a public health emergency or infection being established. The details would require careful thought and the legislation must seek to balance appropriately the interests of the involved parties, but if clinical xenotransplants are to be performed the criminal law must be used to support

94 Noted in, e.g., UNAIDS, *Criminal Law, Public Health and HIV Transmission: A Policy Options Paper* (Geneva: UNAIDS, 2002).

95 R. Martin, 'The Role of Law in Public Health' in A. Dawson (ed.), *The Philosophy of Public Health* (Aldershot: Ashgate, 2009), pp. 11, 19–20.

96 Notable exceptions being the Human Fertilisation and Embryology Acts 1990 and 2008.

97 Specific xeno-legislation exists in, e.g., the Netherlands (Law of 16 May 2002 amending the Law on Special Medical Procedures with Respect to the Introduction of a Prohibition on Xenotransplantation) and has been recommended in, e.g., Norway (I. L. Gjørv, 'Political Considerations of Controversial Medical Issues: Xenotransplantation and Society. A Presentation of the Work of the Norwegian National Working Group on Xenotransplantation' (2004) 45 (Suppl 1) *Acta Veterinaria Scandinavica* S53) and Sweden (Swedish Committee on Xenotransplantation, *From One Species to Another – Transplantation from Animals to Humans: Summary and Statutory Proposals*, Swedish Government Official Report No. 1999: 120 (Stockholm: Swedish Committee on Xenotransplantation, 1999).

early, *anticipatory* action.[98] Such 'anticipatory' actions, where an individual's behaviour creates a risk of harm to others, are in evidence in the criminal law in statutes such as the Firearms Act 1968 (possession of firearms offences) and the endangerment offence under section 33 of the Health and Safety at Work Act 1974. A ladder of sanctions should be introduced, with prosecution and imprisonment at the apex, and the first missed scheduled surveillance appointment or test, for example, should automatically result in activation of the bottom rung of the ladder, *without* the need for proof of infection. The nature of these sanctions and the circumstances in which they apply must be clearly set out, as 'the crucial language in any law seeking to protect the community from the spread or outbreak of an analogous disease situation will lie in how the statute addresses *potential* risks or threats'.[99] Indeed, if action can only be taken on proof of infection with xenotransplantation, this may be too late. The provisions of the amended 1984 Act may be a useful basis for this statute, but it must give health professionals the power to *compel* relevant parties to comply with the surveillance scheme, with using reasonable force being a possibility. Statutory support for compulsory isolation and quarantine in clearly specified circumstances should thus be included, as this precaution is needed to protect individual and public health. With these enforcement powers, the freedom of those who have sought to benefit from a xenotransplant and their obligations to protect others from the risks they, individually, have taken should, as far as possible, be balanced.

The introduction of such a statute is defensible, although its provisions may be controversial, because 'the public expects the state to intervene to prevent known threats to health. The failure of the state to intervene suggests that the threat is minimal, imaginary, or unimportant.'[100] Additionally, Brazier and Harris have argued that '*[i]f* the criminal law is to play an effective role in the control of disease, the law should seek to punish an offender's fundamental irresponsibility rather than focus exclusively on the outcome of such irresponsibility. Society's interest is in deterring infection and the infectious from unjustifiable *risks* to others.'[101] The criminal law should thus be utilised against those who deliberately or recklessly endanger others.[102] Where a xeno-recipient, for example, has been informed of the risk of disease

[98] On anticipatory action, see M. Brazier and J. Harris, 'Public Health and Private Lives' (1996) 4 *Medical Law Review*, 171, 187–89.
[99] Holland, 'The "Catch-22"', 181 (original emphasis).
[100] R. Coker and R. Martin, 'Introduction: The Importance of Law for Public Health Policy and Practice' (2006) 120 Suppl. 1 *Public Health*, 2, 5.
[101] Brazier and Harris, 'Public Health', 182 (original emphasis).
[102] Ibid., 190.

transmission and nevertheless fails to comply with the surveillance regime, it is hard to argue that they do not fall within this category. The criminal law *should* apply to those who 'endanger others by means of reckless conduct risking transmission of disease', because '[t]he core of the wrong done by a person recklessly exposing another to disease is unrelated to the disease.'[103] Indeed, as Coggon argues, 'some imperatives relating to health will best be dealt with by the state, and most defensibly and ethically instituted using criminal regulation'.[104] Legislation is also necessary because civil liberties should not be restricted without such support, and, of course, any statute in England and Wales has to comply with the Human Rights Act 1998.

If it is difficult to clearly limit the scope of these powers in an Act, then it must be asked whether solid organ clinical xenotransplants should be performed. If this hurdle is crossed, then attention must turn to whether there are situations in which it is legitimate to *impose* obligations on those who take risks with the health of others, whilst trying to benefit themselves. As it is currently impossible to determine the nature and magnitude of xeno-risks, recipients must accept that along with the potential benefits, they must also bear the burdens and obligations that are necessary to protect themselves and others. The latter *must* be supported by the criminal law if it is decided that it is appropriate to reap the possible benefits of xenotransplantation, and, in so doing, threaten public health. I agree that '[u]se of the criminal law demands caution because of its potential harshness, but must be seen as a good and proper regulatory mechanism in some instances', [105] and argue that genetically engineered solid organ xenotransplants *are* such an instance.

[103] Ibid. [104] See Coggon in this volume. [105] Ibid.

10 The criminal law and enhancement: none of the law's business?

Nishat Hyder and John Harris

Introduction

From spectacles to telescopes, windmills to watches, bicycles to aeroplanes, vaccinations and medicines, to assisted reproduction and synthetic life, and from writing and printing to e-books and computers, the history of humankind has been one of exploration, discovery and innovation. Whether the outcome is to see further, travel faster or resist disease, we humans have welcomed advances in technology, not least because they defy the limits imposed on us by nature and enhance our lives. Today, to add to the list above, there are drugs available that can help focus attention and manipulate information, sharpen memory and promote wakefulness – in other words, enhance many dimensions of cognitive function. Chemical cognitive enhancers (CCEs) refer to drugs that were often originally developed to treat those with cognitive disabilities or neuropsychiatric disorders; however, when used by healthy adults they have been shown to boost brain power.[1]

Given our long history of embracing enhancement tools and technologies, it is odd that CCEs have been regarded with such suspicion, given their respectable safety record in clinical applications. (Methylphenidate (Ritalin) has been prescribed for children since the 1980s for a condition that is not life-threatening or painful – attention deficit disorder.[2]) CCEs are referred to, even by critics and sceptics, as '*enhancement* drugs' or 'smart drugs' – names that advertise their beneficial purpose. Yet the dispensation and use of CCEs for non-therapeutic purposes is principally forbidden in the UK by either the criminal law or professional regulatory standards; or indeed societal norms that seek to inhibit use of CCEs, often referring to their use,

[1] H. Greely, B. Sahakian, J. Harris, R. Kessler, M. Gazzaniga, P. Campbell and M. Farah, 'Towards Responsible Use of Cognitive Enhancing Drugs by the Healthy' (2008) 456 *Nature*, 702.

[2] D.J. Safer, J.M. Zito and E.M. Fine, 'Increased Methylphenidate Usage for Attention Deficit Disorder in the 1990s' (1996) 98(6) *Pediatrics*, 1084.

quite simply, as cheating.[3] In this essay, we explore the role of the criminal law, and indeed other regulative mechanisms, in controlling access to CCEs.

So far, neither the legal nor the social sanctions imposed against CCEs have stopped widespread use of the drugs. Across top-class universities, including Ivy League and Oxbridge colleges, and even within high schools, CCEs are becoming an increasingly popular study aid, rivalling triple-shot cappuccinos and Cliff Notes. The headline 'Drugs on College Campuses' usually evokes images of reckless, hedonistic young adults headed for crash and burn, but 'smart drugs', to use the vernacular, present a very different picture. CCEs are no party drug. Users are motivated by a work, or at least a success, ethic: by a desire to work harder, faster and more effectively, and to balance workload with equally demanding extracurricular/leisure activities. In a world where the pursuit of knowledge and advancement and the acquisition of knowledge and know-how is encouraged as a worthwhile goal, it is striking that the status of CCEs for off-label use by healthy and responsible adults remains legally and socially suspect.

It is true that there is also a strong puritan streak in many 'Western' societies, and tolerance of hedonism and the recreational use of drugs in particular has often been conspicuous by its absence. It is nothing short of genuinely mysterious how much hostility has been directed at those seeking to use chemical means of enhancement of knowledge, know-how and indeed physical and athletic performance.

We are not principally concerned here to put forward the positive case for the use of CCEs by healthy adults – that has been done elsewhere;[4] rather, we intend to explore the issue of potentially regulating CCEs. That is to say, we address the question of whether and to what extent regulating CCEs via the criminal law is appropriate, and, if so, what sort of regulatory framework and strategy would best address issues of access to and use of CCEs.

3 Such responses have been well documented in the media; see, for example, M. Szalavitz, 'Popping Smart Pills: The Case for Cognitive Enhancement', *Time Magazine*, 6 January 2009 (see www.time.com/time/health/article/0,8599,1869435,00.html); A. Frean and P. Foster, 'Cheating Students Turn to Smart Drug for Edge in Exams', *The Times Online*, 23 June 2007 (see www.timesonline.co.uk/tol/news/uk/education/article1975271.ece); 'Sports-Style Dope Tests to Weed Out Drug-Cheat Students', *Mail Online*, 1 October 2009 (see www.dailymail.co.uk/news/article-1217274/Doping-tests-exam-students-weed-drug-cheats.html); 'Exam-Boosting Drug Tests Loom', BBC News, 21 May 2008 (see http://news.bbc.co.uk/1/hi/health/7412719.stm).

4 See J. Harris, *Enhancing Evolution: The Ethical Case for Making Better People* (Princeton and Oxford: Princeton University Press, 2007) and Greely *et al.*, 'Towards Responsible Use'.

The drugs

There are a number of drugs currently available that can aid cognitive function in healthy adults; in this essay we focus on the more prevalent varieties. Stimulants such as methylphenidate (better known as Ritalin) and mixed amphetamine salts (Adderall) are popular choices of CCE. Both drugs are used to treat attention deficit hyperactivity disorder (ADHD) in children. Modafinil (marketed under the name Provigil or Alertec) is another stimulant commonly used for enhancement purposes, although it was originally developed to treat narcolepsy. Finally, drugs developed to help sufferers of Alzheimer's disease, such as donezepil (marketed as Aricept), are also exploited as CCEs.[5] Research suggests that stimulants are generally preferred for enhancement purposes over drugs such Aricept,[6] although it might simply be the case that they are better known as a result of popular media coverage, and, arguably, easier to access.

There is certainly clear evidence that indicates these drugs can have a positive impact on various aspects of cognition. Results from a study by the University of Cambridge on the impact of modafinil in healthy adults have shown improvements in spatial planning, response inhibition, short-term memory and visual memory.[7] Atomoxetine too (used to treat ADHD) has been shown to improve response inhibition in healthy adults.[8] Furthermore, a plethora of anecdotal evidence from users of various CCEs points to properties of cognitive enhancement; users report heightened alertness, concentration, mental stamina and even motivation, resulting in increased productivity.[9]

[5] See C. R. Housden, S. Morein-Zamir and B. J. Sahakian, 'Cognitive Enhancing Drugs: Neuroscience and Society', in J. Savulescu, R. ter Meulen and G. Kahane (eds.), *Enhancing Human Capacities* (Oxford: Wiley-Blackwell, 2011), Chapter 7; P. Bee, 'Smart Drugs for Straight As', *The Times Online*, 14 May 2007 (see www.timesonline.co.uk/tol/life_and_style/health/features/article1780141.ece); 'Smart Drugs Boost Pilot Memory', BBC News, 8 July 2002 (see http://news.bbc.co.uk/1/hi/health/2116476.stm).

[6] B. Maher, 'Poll Results: Look Who's Doping' (2008) 425 *Nature*, 674.

[7] D. C. Turner, T. W. Robbins, L. Clark, A. R. Aron, J. Dowson and B. J. Sahakian, 'Cognitive Enhancing Effects of Modafinil in Healthy Volunteers' (2003) 165 *Psychopharmacology (Berlin)*, 260.

[8] S. R. Chamberlain, A. Hampshire, U. Muller, K. Rubia, N. Del Campo and K. Craig, 'Atomoxetine Modulates Right Inferior Frontal Activation During Inhibitory Control: A Pharmacological Function Magnetic Resonance Imaging Study' (2009) 65 *Biological Psychiatry*, 550.

[9] M. Talbot, 'Brain Gain: The Underground World of Neuroenhancing Drugs', *The New Yorker*, 27 April 2009; N. Lennard, 'One in Ten Takes Drugs to Study', *Varsity* (Issue 693), 6 March 2009; B. J. Sahakian and S. Morein-Zamir, 'Professor's Little Helper' (2007) 450 *Nature*, 1157.

Crucially, it is important to understand that these drugs do not affect the quality of cognition itself; they do not produce deeper insights or enhance originality. Rather, the drugs are tools that can aid cognitive function: they manipulate neurotransmitters found in the brain (the chemicals that modulate cognition) and this process results in measurable or tangible differences in, say, attention or memory. Thus, these drugs might legitimately be compared to acceptable forms of enhancement, such as caffeine, private tutors, sleep, nutrition, exercise and hard work – all of which can lead to the positive effects enumerated by CCE users. In addition, it is important to note that the various domains of cognition are controlled by different systems within the brain, and none of the CCEs developed to date are comprehensive in their ability to enhance cognition; so, at present, the impact of any particular drug will be limited.[10]

The use of CCEs for non-therapeutic reasons has attracted considerable criticism. One of the main criticisms is the issue of safety: these drugs have not been through the rigorous clinical trial process for enhancement purposes. Research trials and anecdotal evidence have pointed to various side effects. For instance, side effects of using stimulants include headaches, nausea, anxiety, irritability, elevated blood pressure, cardiovascular complications, skin reactions, poor bladder control, diarrhoea, loss of appetite, experiencing a 'low' as the drug wears off, dependence and impaired social interaction.[11] The incidence of these side effects is inconsistent, and some users report none at all. Further, as many users self-medicate, they will be unaware of contraindications that may cause side effects and possibly interfere with the effectiveness of the drug. This does compromise the validity of such anecdotal evidence against CCEs, but also highlights the importance of further research and information dissemination. A further argument in favour of caution is that although there is evidence of immediate effects of these drugs on healthy adults, the long-term effects are unknown.

The list of possible side effects above may seem forbidding, but read the leaflet on almost any over-the-counter medication and not dissimilar daunting lists of side effects can be found. For example, the list of possible side effects of using ibuprofen includes nausea, vomiting, diarrhoea, indigestion, abdominal pain, headache, dizziness, fluid retention, raised blood pressure, gastritis, duodenal or gastric ulcers, allergic reactions and bronchospasm.[12] That said, the possibility of side effects is a

[10] Housden *et al.*, 'Cognitive Enhancing Drugs'.
[11] Ibid.; see also Talbot, 'Brain Gain' and Lennard, 'One in Ten'.
[12] See www.nhs.uk/Conditions/Painkillers-ibuprofen/Pages/Side-effects.aspx.

risk associated with accepting any medical treatment, and the bare fact of there being side effects is not necessarily enough to preclude the use of a particular treatment. For example, certain drugs prescribed as an adjunctive treatment for psychotic disorders[13] (for example, Clonazepam, which is a benzodiazepine derivative) are highly addictive; however, their therapeutic benefit outweighs the side effect of possible addiction.[14] Drugs used to treat depression, such as the selective serotonin reuptake inhibitor (SSRI) Citalopram, have unpleasant initial side effects, such as nausea, dizziness, sleepiness and weakness, which are outweighed by their eventual therapeutic utility.[15] Further research is required into both the short- and long-term effects of using CCEs in order to enable users to make an informed decision.

The effectiveness of any CCE will depend on the user. As individuals we will all have a different natural baseline level of chemicals in our brain. The 'right' level of chemicals present will result in optimum cognitive function. Thus, determining the correct dose is critical to achieving optimum levels of performance. Because we all have different baselines, some individuals will be functioning at a higher level than others, and studies have shown that the correlation between dosage and effectiveness of some CCEs is not linear, but an 'inverted U' effect. That is to say, too high a dose will result in impaired cognitive function, explaining why high-functioning individuals can experience few, no or even adverse effects on cognition when taking CCEs.[16] It is possible that a 'ceiling effect' inherent in the design of the task used during the trial, or perhaps in the drug itself, is accountable for CCEs failing to take effect in some individuals; but this theory does not explain why the performance of some individuals deteriorates.[17]

[13] D. Gillies, A. Beck, A. McCloud, J. Rathbone and D. Gillies, 'Benzodiazepines Alone or in Combination with Antipsychotic Drugs for Acute Psychosis' (2005) 4 *Cochrane Database of Systematic Reviews*, CD003079; F. Curtin and P. Schulz, 'Clonazepam and Lorazepam in Acute Mania: A Bayesian Meta-analysis' (2004) 78(3) *Journal of Affective Disorders*, 201.

[14] J. Riss, J. Cloyd, J. Gates and S. Collins, 'Benzodiazepines in Epilepsy: Pharmacology and Pharmacokinetics' (2008) 118(2) *Acta Neurologica Scandinavica*, 69. See also www.nlm.nih.gov/medlineplus/druginfo/meds/a682279.html.

[15] See www.patient.co.uk/medicine/Citalopram.htm.

[16] M. A. Mehta, A. M. Owen, B. J. Sahakian, N. Mavaddat, J. D. Pickard and T. W. Robbins, 'Methylphenidate Enhances Working Memory by Modulating Discrete Frontal and Parietal Lobe Regions in the Human Brain' (2002) 20 *Journal of Neuroscience*, RC65; M. A. Mehta, R. Swainson, A. D. Ogilvie, B. J. Sahakian and T. W. Robbins, 'Improved Short-Term Spatial Memory but Impaired Reversal Learning Following the Dopamine D(2) Agonist Bromocriptine in Human Volunteers' (2001) 159 *Psychopharmacology (Berlin)*, 10.

[17] Housden *et al.*, 'Cognitive Enhancing Drugs'.

Where an individual sits on the inverted U is dependent on a number of factors, including genotype.[18]

There are 'trade-offs' of taking CCEs to consider. For example, a drug may have simultaneously linear and inverted U effects on two different areas of cognition (this has been found with methylphenidate).[19] The human brain is finely tuned; the chemical dopamine is involved in a multitude of functions, so manipulating it to enhance memory function may cause another neurological side effect. CCEs have also been criticised for compromising creativity or social interaction. One study focusing on the potential impact of Adderall on creativity suggests that the drug does not necessarily impact creativity. The results point to an inverted-U pattern: whilst on the drug, highly creative individuals became less creative and vice versa.[20] We are still in the early stages of developing and understanding cognitive enhancers and it is impossible to make any definitive statement on the effects and side effects of these drugs. Indeed, despite its widespread use, exactly how modafinil operates is not yet fully understood.[21]

The CCEs that are currently available are by no means perfect. However, the development of therapies to aid cognition is highly lucrative within the public and private sectors. Research into developing treatments for neurodegenerative conditions such as Alzheimer's disease is ongoing, and results in that area may well translate into cognitive enhancement proper. Our understanding of the complexities of the human brain and cognition is just beginning; as our understanding increases, so too will our ability to create more effective drugs.

The law

CCEs have not been specifically targeted within the law or an alternative regulatory process. Despite studies indicating significant usage of these drugs within schools and universities[22] – and there is no reason to believe

[18] C. M. Diaz-Asper, D. R. Weinberger and T. E. Goldberg, 'Catechol-O-Methyltransferase Polymorphisms and Some Implications for Cognitive Therapeutics' (2006) 3 *NeuroRx*, 97.

[19] K. Konrad, T. Gunther, C. Hanisch and B. Herpertz-Dahlmann, 'Differential Effects of Methylphenidate on Attentional Functions in Children with Attention-Deficit/Hyperactivity Disorder' (2004) 43 *Journal of the American Academy of Child and Adolescent Psychiatry*, 191.

[20] M. J. Farah, C. Haimm, G. Sankoorikal, M. E. Smith and A. Chatterjee, 'When We Enhance Cognition with Adderall, Do We Sacrifice Creativity? A Preliminary Study' (2009) 202 *Psychopharmacology (Berlin)*, 541.

[21] Housden *et al.*, 'Cognitive Enhancing Drugs'.

[22] Talbot, 'Brain Gain'; Lennard, 'One in Ten'; Sahakian and Morein-Zamir, 'Professor's Little Helper'.

that usage has not spread to professional circles – there is no trend towards sector-specific regulation or codes of conduct relating to the use of CCEs. At present, regulation bearing on CCEs arises mainly from the Medicines Act 1968 and the Misuse of Drugs Act 1971, which we shall examine briefly now.

The CCEs we have discussed so far in this essay are prescription-only drugs, the sale and supply of which are regulated by section 58 of the Medicines Act:

Section 58 Medicinal products on prescription only.

[. . .]

(2) Subject to the following provisions of this section–
(a) no person shall sell by retail, or supply in circumstances corresponding to retail sale, a medicinal product of a description, or falling within a class, specified in an order under this section except in accordance with a prescription given by an appropriate practitioner; and
(b) no person shall administer (otherwise than to himself) any such medicinal product unless he is an appropriate practitioner or a person acting in accordance with the directions of an appropriate practitioner.
(3) Subsection (2)(a) of this section shall not apply–
(a) to the sale or supply of a medicinal product to a patient of his by a doctor or dentist who is an appropriate practitioner, or
(b) to the sale or supply of a medicinal product, for administration to an animal or herd under his care, by a veterinary surgeon or veterinary practitioner who is an appropriate practitioner.

This latter exception is truly exceptional. It might be taken to imply that while the creation of smarter animals is, if not acceptable, at least permitted, smarter people are a problem so serious as to merit legal sanction. Of course, it is probable that the drafters of the Medicines Act did not consider the possibility of herdsmen or farmers bent on smartening up their animals (except for agricultural shows, of course). Recently, however, serious consideration is being given to the possibility of both embryos and mature creatures being created containing human and animal admixtures, and there is good reason to question whether the results of such consideration will be as sanguine as to the supply of enhancers to animals.[23]

[23] See, for example, The Working Group Chaired by Sir David Weatherall FRS FMedSci, *The Use of Non-Human Primates in Research* (London: The Academy of Medical Sciences, The Medical Research Council, The Royal Society and The Wellcome

The Act clearly states that the sale, supply and administration of prescription-only drugs are subject to control; however, it remains silent on the issue of possession of prescription-only drugs. Thus, a person who, without an appropriate prescription, purchases and uses a prescription-only CCE will not be subject to any legal sanction or reprimand, whereas depending on their geographical location, the vendor may well be charged.

However, certain prescription-only drugs are also classified as controlled substances under the Misuse of Drugs Act 1971, and therefore regulated with considerably more rigour. Not only is the production and supply of controlled substances prohibited (section 4), but so too is possession (section 5), as well as a number of related offences. Schedule II of the Misuse of Drugs Act lists the substances controlled under the Act and their classification. Under this, methylphenidate (Ritalin) and amphetamines (Adderall) are both Class B drugs; unlawful possession could lead to up to five years' imprisonment and/or an unlimited fine, and dealing could lead to up to fourteen years' imprisonment and/or an unlimited fine. The Act does distinguish between lawful and unlawful possession of drugs that are prescription-only, so a Ritalin or Adderall user would have to produce a valid prescription to avoid being charged with a criminal offence.

That, in a nutshell, is the purview of the two Acts in relation to CCEs, and the resulting legal position is both odd and inconsistent. Not all CCEs are classified as drugs under Schedule II of the Misuse of Drugs Act – and there appears to be no discernible reason as to why certain CCEs are classified and more harshly regulated, and others not, when the use of any or all such drugs for enhancement is questionable. It is interesting to note that a drug such as Ritalin, which has been extensively tested, approved and used in the treatment of children for many years, is subject to strict regulation under the Misuse of Drugs Act, whereas modafinil (about which much less is known), albeit a prescription-only drug, is not.

So far we have established that CCEs can be legally obtained and used, either by prescription or by buying a non-classified enhancement drug from a legal (this will invariably mean an overseas and/or Internet) seller. This latter method is fairly straightforward; however, the former

Trust, December 2006); The Academy of Medical Sciences, *Inter-Species Embryos* (London: *The Academy of Medical Sciences*, June 2007). The Academy of Medical Sciences currently has a working group addressing issues concerning *Animals Containing Human Material*. John Harris was, and in the latter case still is, a member of the working parties involved in each of these studies.

warrants further discussion. It is important to be aware of the context in which the use of CCEs came about. As mentioned above, these drugs were originally developed as therapies for specific conditions – their non-therapeutic enhancement properties came to light subsequently. Consequently, they have only been through the clinical trial process and licensed for certain, therapeutic purposes. Notwithstanding this, doctors can and do prescribe for 'off-label' purposes as part of common medical practice.

The General Medical Council (GMC) has issued good practice guidance for off-label prescriptions:

> 20 When prescribing a medicine for use outside the terms of its licence you must:
>
> (a) be satisfied that it would better serve the patient's needs than an appropriately licensed alternative
> (b) be satisfied that there is a sufficient evidence base or experience of using the medicine to demonstrate its safety and efficacy; the manufacturer's information may be of limited help, in which case the necessary information must be sought from other sources
> (c) take responsibility for prescribing the medicine and for overseeing the patient's care, monitoring and any follow up treatment, or arrange for another doctor to do so (see also paragraphs 25–27 on prescribing for hospital outpatients)
> (d) make a clear, accurate and legible record of all medicines prescribed and, where you are not following common practice, your reasons for prescribing the medicine in the patient's notes.[24]

The GMC guidelines emphasise the need for the doctor to take clinical responsibility when prescribing off-label. And, whilst the patient's best interests are paramount, there is no explicit requirement for those interests to be strictly therapeutic. Indeed, there are many treatments that require medical oversight that are not based on strict clinical need, such as prescribing beta-blockers to cope with work or exam-related stress and anxiety, or cosmetic or other 'elective' forms of surgery.

Provided one can find a doctor willing to exercise his/her discretion to prescribe certain drugs for off-label purposes, CCEs are legally obtainable. Under this scenario, although the doctor remains responsible for his/her clinical decision, questions of liability and professional conduct would only arise upon facts occurring that give rise to a civil

[24] GMC Supplementary Guidance, *Good Practice in Prescribing* (September 2008) (see www.gmc-uk.org/static/documents/content/Good_Practice_in_Prescribing_Medicines_0911.pdf).

cause of action. Notably, the pharmaceutical companies are prohibited, by virtue of Article 87(1) of the Directive 2001/83,[25] from advertising off-label uses for the products they develop. Although the Directive distinguishes between advertising and providing information, the distinction is particularly fine, and tension remains between the position of the doctor – who is at liberty to prescribe off-label products – and the pharmaceutical company – which is prevented from advertising its products for purposes it knows the products will be used for regardless.

The law as it stands in relation to CCEs is riddled with discrepancies and inconsistencies. Whilst some lawyers are complacent about inconsistencies in and between laws, it is obvious in logic that a proposition cannot simultaneously be both right and wrong. First, the legality of a purchaser/user's action is entirely arbitrary; it hinges on the choice of CCE rather than the fact of the drug being a CCE, or indeed on the seriousness of associated risks or side effects, as well as (in some instances) the choice of prescriber. Second, the law is geared towards targeting the seller, supplier or producer, as opposed to the purchaser/user, which sets a somewhat confusing standard. Whether or not one agrees with the use of CCEs for non-therapeutic purposes, it is clear that targeted reform within the law is required to elucidate and make consistent the legal position, and to provide a consistent rationale one way or the other.

Some might be inclined to argue that, safety risks aside, CCEs hold the ability to alter mankind through their enhancing function and that is clearly sufficiently serious as to merit quite strict rules governing their use and availability. However, we know that acquiring knowledge alters the physical structure of the brain,[26] and that food and exercise do the same for the body, yet this dramatic 'alteration of mankind' is, so far as we are aware, nowhere taken to license the interference of the criminal law. What is needed is some reference to the harm or wrongfulness of effecting such alterations.

[25] Article 87(1) of Directive 2001/83/EC (Community Code relating to Medicinal Products for Human Use) states that: 'Member States shall prohibit any advertising of a medicinal product in respect of which a marketing authorization has not been granted in accordance with Community law.'

[26] In 2003, a group from University College London showed that the volume of grey matter in the posterior hippocampus of London taxi drivers increases in size due to the navigational expertise and knowledge that they acquire through their jobs (E. A. Maguire, H. J. Spiers, C. D. Good, T. Hartley, R. S. J. Frackowiak and N. Burgess, 'Navigation Expertise and the Human Hippocampus: A Structural Brain Imaging Analysis' (2003) 13 *Hippocampus*, 208).

Regulatory reform and strategy

So, why regulate? At present, there enough gaps in the law to allow most of those who want to use CCEs to do so without threat of legal repercussions. Meanwhile, policy-makers can maintain a diplomatic silence on the issue, avoiding a messy public debate and potential loss of political capital. However, even our brief exposition of the rate of development in CCEs, rates of use and current legal position indicate that simply maintaining the status quo is deeply unsatisfactory for all parties concerned, and, moreover, irresponsible. If questions are to be raised over whether or not CCEs should be used, and, if so, which ones and by whom, clear and defensible policy and regulatory strategy are essential.

In recent years, government has begun to take a deeper interest in issues of public health and policy. Indeed, the scope of the term 'public health' has widened incrementally, from basic sanitation and nationwide vaccination schemes, to efforts to control smoking and excessive alcohol consumption, and, most recently, the encouragement of diet and exercise. Recent governments have spent considerable time, money and energy on advocating healthy lifestyle choices through sweeping national campaigns and local initiatives. Their message is crystal clear: drink two litres of water daily, exercise three to four times a week, eat five portions of fruit and vegetables every day, sleep for eight hours every night and you *will* be a glowing, paragon of health – and a responsible citizen to boot.

The lifestyle adjustments that recent governments have so vocally encouraged are all forms of enhancement. For example, it has been proven that regular exercise has significant physical and mental health benefits, as well as cosmetic and indeed social acceptance gains for the individual. Exercise is also responsible for observable physical and physiological changes to the organism. Exercise is not compulsory, but exercising will undeniably result in an enhancement of one's health and wellbeing, and the individual benefits are ultimately reflected at the societal as well as the personal level. A healthier population means less strain on the National Health Service, fewer sick days for employers to account for, greater productivity and therefore fiscal savings across several sectors. Analogously, a 'smarter' population would be likely to have equal, if not more, gain. Education would be more efficient because students would process knowledge more effectively, have enhanced memory, organisational skills, and so forth. Of course, this is a hypothesis that would need to be tested, but it certainly cannot be tested while the methods of doing so are illegal. Indeed, taking CCEs might well come to be seen as no different to going to the gym.

We are encouraged to lead better lives; not only to work, but to work hard, a standard that is increasingly difficult to maintain given the twenty-four-hour work-day culture. CCEs will enable people to work at a higher capacity, with greater focus, speed and skill, all of which will translate into a more productive and efficient workforce, and, further, a working population able to finally balance work and private life, thereby reducing stress and ill health. Not only are these effects intrinsically and instrumentally valuable for individuals and for society, but, again, there are significant resource savings to be captured. Given the government trend towards promoting certain enhancements, and conversely discouraging harmful behaviour (such as smoking and excessive alcohol consumption), it is not clear why CCEs should be distinguished as an unacceptable enhancement.

It will be clear to the reader that the authors of this essay are principally in favour of CCEs, and bearing this in mind we now turn to discussing issues relating to access and use of CCEs.

Access

Who should have access to CCEs? One of the main questions in the enhancement debate is whether access to CCEs should be universal, limited or controlled and monitored. Should it be the reserve of educational institutes? Or those with particularly high-responsibility jobs and onerous working hours? At present, pilots, certain shift workers and those in the Armed Forces are allowed, indeed often encouraged, to take CCEs, by virtue of the nature of their employment.[27] Evidence is hard to come by, particularly in a climate of illegality, but it is widely believed that US forces are provided with amphetamines and alertness drugs for combat missions.[28] Whilst it is understandable why these professions have been singled out, there are many other professions with an arguably equal claim to use CCEs. What of the medical profession? Junior doctors who work notoriously lengthy shifts? What about surgeons required in the operating theatre for hours during complex procedures? How about emergency and relief workers in the aftermath of a disaster? Those with jobs in the City work infamously difficult hours – why not allow them access to CCEs too? The question is, would it matter who had access to these drugs?

[27] P. Roxby, 'Do "Smart Drugs" Really Make Us Brainier?', BBC News, 3 April 2011 (see www.bbc.co.uk/news/health-12922451).

[28] See S. Rose, 'Brain Gain?' in P. Miller and J. Wilson (eds.), *From Smart Pills to Designer Babies and Extended Lifespans, Technology Now Promises to Transform our Very Nature. Better Humans* (London: Demos, 2006).

The answers to these questions bear crucially on the legitimacy or wisdom of controlling access to CCEs via the criminal law and in other ways. Analogous arguments have been made for the use of beta-blockers by professional and amateur musicians. This is still controversial, but perhaps the degree of controversy is in this case, as in so many others, evidence of insufficient consensus to justify legislation.[29]

Regulating access to CCEs on the basis of employment type will necessitate making a certain value judgment on the 'worth' of a job, an exercise that is extremely difficult, as well as notoriously invidious. Why should the need of, say, a single, junior doctor, on track to become a consultant oncologist, be greater than that of a single mother, who works two different jobs in retail to support her young family and takes a class in Russian literature during the evenings? A further question related to employment-dependent access is whether, if certain jobs guaranteed access to CCEs, employers could make employment conditional on taking CCEs? Would a condition of that sort infringe autonomy in an unsustainable fashion? Or could it be justified as just another burdensome clause in an employment contract, in the same vein as working 'reasonable hours' outside standard hours of business?

We submit that voluntariness is crucial and measures ought to be built into any regulatory framework to protect people against forced use of these drugs. Although we contend that the benefits of CCEs are significant, like many decisions in life, taking them ought to remain a free choice for the individual. Contemporary UK jurisprudence has a strongly developed doctrine of autonomy, enshrining – to a point – control over one's body and lifestyle. These drugs are not akin to staying

[29] For media coverage on musicians' use of beta-blockers and the controversies arising therefrom, see V. L. Watts, 'Beta-Blockers Used by Musicians, Athletes, Students to Enhance Performance', *The Inquirer*, 16 August 2010 (see http://articles.philly.com/2010-08-16/news/24973169_1_beta-blockers-graduate-student-performance-anxiety); C. Elliott, 'In Defense of the Beta Blocker', *The Atlantic*, August 2008 (see www.theatlantic.com/magazine/archive/2008/08/in-defense-of-the-beta-blocker/6961); B. Tindall, 'Just One More to Calm the Nerves . . .', *The Guardian*, 5 June 2008 (see www.guardian.co.uk/music/2008/jun/05/classicalmusicandopera.news); B. Tindall, 'Better Playing Through Chemistry', *The New York Times*, 17 October 2004 (see www.nytimes.com/2004/10/17/arts/music/17tind.html). See also 'Beta Blockers in Situational Anxiety' (1985) 326 (8448) *The Lancet*, 193; I. M. James, R. M. Pearson, D. N. W Griffith and P. Newbury, 'The Effect of Oxprenolol on Stage Fright in Musicians' (1977) 310(8045) *The Lancet*, 952; C. O. Brantigan, T. A. Brantigan and N. Joseph, 'Effects of Beta Blockers and Beta Stimulation on Stage Fright' (1982) 72 *American Journal of Medicine*, 88; K. A. Neftel, R. H. Adler, L. Käppelli, M. Rossi, M. Dolder, H. E. Käser, H. H. Bruggesser and H. Vorkauf, 'Stage Fright in Musicians. A Model Illustrating the Effect of Beta Blockers' (1982) 44(5) *Psychosomatic Medicine*, 461; I. M. James, W. Burgoyne and I. T. Savage, 'Effect of Pindolol on Stress-Related Disturbances of Musical Performance: Preliminary Communication' (1983) 76(3) *Journal of the Royal Society of Medicine*, 194.

late in the office during a busy period or working odd hours to take a conference call on the other side of the world; they have a definite physiological effect that can only be controlled insofar as an individual chooses to take the drugs – or not. Forced use of CCEs would not only contravene established common law principles, there would also be a strong case for breach of the European Convention rights, now enshrined in the Human Rights Act 1998.

We recognise, of course, that social, peer group and indeed financial pressures militate against full voluntariness in choices such as these. However, such pressures attend in greater or lesser degree the vast majority of personal, work, lifestyle and indeed moral choices that we all make. Morality and ethics (the study of morality), and indeed the law itself, may be thought of as mechanisms of social and individual control.

It is our position that enhancements ought not to be limited to people on the basis of arbitrary factors, such as job title. We have argued that CCEs can be of immense benefit to individuals and society, and thus it would be foolish and self-defeating to restrict access to something that is manifestly a personal and a public good. The motivation behind using CCEs is of little consequence: it matters not whether someone uses CCEs to write a paper, perform surgery, read Joyce's *Ulysses* or organise their extensive CD collection. What does matter is that everyone has equal or at least fair and reasonable access to CCEs if they so wish. However, these drugs can have short-term side effects and, in some cases, little is known of potential long-term side effects. In short, there is risk involved. Nonetheless, the risk is not so great that a competent adult[30] could not voluntarily consent to that risk in much the same manner as we assume risks when we cross the road, take medicine or drink one glass too many of champagne at a party.

Monitoring

Given that there are radically insufficient reasons to restrict access to CCEs, the question then turns to whether use should be monitored. Legal access to certain CCEs is guarded by the medical profession – but are doctors the appropriate gatekeepers to smart drugs? In fact, should access be medically guarded at all? When Ritalin is prescribed to a child for the purpose of treating ADHD, clinical oversight is, of course, entirely appropriate. However, Ritalin, when used (by an adult) as an

[30] This essay is only concerned with the use of CCEs by adults. Arguments pertaining to using CCEs on children deserve specific attention, but are outside the scope of this essay.

enhancement drug, is actually of very little clinical interest. As mentioned previously in this essay, health professionals are often involved in giving treatment for non-therapeutic purposes; while certain non-therapeutic procedures (for instance, cosmetic surgery) require the skill of trained medical practitioners, others do not.

When CCEs are used for purely non-therapeutic purposes, there is a strong argument in favour of effectively 'de-medicalising' the use of these drugs. After all, one would not consult a doctor prior to drinking several litres of a caffeinated drink, taking Pro Plus, various herbal remedies, or hiring a private tutor – using CCEs is not in principle, or indeed in practice, sufficiently different to justify the increased scrutiny or the intervention of the law. In fact, access to various arguably non-therapeutic treatments has slowly been relaxed over time. Drugs such as the contraceptive pill and Viagra were once subject to prescription-only status, but are now more freely available.[31] Whether this is simply a result of growing confidence in the treatment over a period of time, or a retraction of an overly protective and paternalistic measure, the trend merits consideration.

Central to this discussion is the role of the health professional. That is to say, is it appropriate to broaden the role of the GP to cover dispensing enhancement drugs? Moreover, if universal access is the preferred strategic route, is it even practicable? Further trials will of course be necessary, in order to determine the optimum dosage for enhancement purposes, but once this is clarified a medical barrier seems redundant. Indeed, it is somewhat patronising to have doctors guard access to non-therapeutic drugs, especially as they have no control over access to the many therapeutic drugs available over the counter. From the point of view of the clinician, the idea of doling out study aids might appear outside the scope of their job description – and rightly so. If a user suffered an adverse reaction to these drugs, they would of course be entitled to medical attention. Doctors are, we believe, duty-bound to provide care, regardless of the circumstances in which the harm arose, despite increasing attempts to erode this admirable good Samaritan ethic.

Although defining the role of a doctor based on a distinction between therapeutic and non-therapeutic treatment collapses, distinguishing on

[31] Viagra is now available over the counter, and the contraceptive pill is obtainable at contraceptive clinics, as well as through GP services and pharmacies. There have been a number of pilot schemes throughout the UK, making the pill available over the counter in order to target specific demographic groups, and some pharmacies offer online supply services (see, for example, https://onlinedoctor.lloydspharmacy.com/womens_health/the_pill/repeat_of_my_current_pill?gclid=CMSY-Y2OrqkCFcod4QodtGnTLQ). The emergency contraceptive 'morning-after' pill is available over the counter.

the basis of the requirement for professional skill and care does not. If you develop symptoms of cold or flu, you are expected to visit your local pharmacy, purchase the appropriate treatments and 'self-medicate' according to the instructions included on the packaging. There is scant evidence to suggest that CCEs should be treated any differently.

At this point, readers will recall the foregoing discussion in this essay on the differing impact and effect of the various types of enhancement drugs on the brain, and the link between dosage and individual baseline. The *optimum* dose for *optimum* functioning of most medicines will differ according to individual variation in physiology, lifestyle, and so on. CCEs are no different. In an ideal world, all medical treatment would be carefully, individually tailored; however, we do not live in an ideal world, and constraints on time and resources call for pragmatism. Medicines that are offered over the counter (and, indeed, even those that are specifically prescribed) are made available in a safe, *effective*, minimum dosage, with clear instructions as to the safe maximum dosage. It is up to the individual to ascertain the appropriate drug, dose and timing. For example, if a person is afflicted with a headache, it is up to the individual to opt, for example, for paracetamol, ibuprofen or aspirin as a treatment, and then, bearing in mind the safe maximum dose limits, elect how, when and how many tablets to take for his/her specific needs. Indeed, most consumers are also aware, for example, of the availability of enhanced effect through using certain drugs in combination – for instance, ibuprofen and paracetamol may be used in combination. Equally, most are aware that advice on combining drugs needs to be sought from professionals. There is no reason why CCEs cannot be made available in the same fashion: a safe, effective minimum dose that will be safe enough to be made generally available.[32] This approach will also facilitate a comprehensive system of access and individual choice.

We have advocated universal adult access to CCEs; however, a meaningful principle of access requires considering a fair pricing strategy and income levels across all demographic groups. If CCEs were to enter the marketplace, they would do so at a price determined by market forces (recuperating production costs, creating profit, product demand, and so forth) – which may exclude purchasers with a low income, or no income

[32] Determining a safe and effective minimum dose will involve collecting and analysing data on the effects of chronic use of CCEs (by healthy adults), which will inform what constitutes a long-term effective dose and long-term side effects. Although discussion on this particular issue is beyond the scope of this essay, it is important to note that a reassessment of the current regulatory framework in place by the European Medicines Agency might be necessary in order to facilitate an appropriate and informed analysis of the long-term use and effects of CCEs.

at all. Of course, many things – including enhancements – cost money, and just because something is an enhancement does not mean we can demand free, universal access. Egalitarian arguments of fairness and justice are valid in this context, but ideals must be balanced against the limits of the public purse.

The government subsidises many products and services to those who cannot otherwise afford to pay for them. To refer to an earlier example, gym membership is (often) available at a subsidised rate in local council facilities for those in full-time education or on income support. There is no reason why access to CCEs cannot also be means-tested and subsidised accordingly, in order to answer criticisms of equal and fair access. 'De-medicalising' CCEs opens up the range of regulatory methods: these drugs could be offered through institutes of higher education or employers (who would in turn be subsidised or compensated by public funding if necessary) or through local public services. Furthermore, the state subsidy need not come out of the NHS budget; if CCEs were to be available as a study aid, for example, it would be appropriate for the education budget to cover the costs of subsidies.

Regulatory strategy on access to and provision of CCEs is an open book at present. There is, without doubt, a place for draconian command-and-control regimes, such as that in place under the Misuse of Drugs Act 1971, which was enacted as a measure against crime and social disorder. The reasons that motivate strict drug policy and criminal sanctions vis-à-vis heroin or cocaine do not apply to CCEs, which are, we argue, beneficial on an individual and societal basis. In considering the case of CCEs, it is our belief that the role of regulation ought to be to enable informed freedom of choice. Regulatory strategy in this instance would be best directed towards focusing on issues such as safety, quality control, advertising standards and redressing the asymmetry of information. The regulatory model proposed here is a liberal one that may not be appropriate to implement without further research. Further, extensive work into short- and long-term effects of CCE use, as well as determining a safe dosage, understanding the interaction of these drugs with other treatments, and the mechanism of the drugs themselves may be prudent before the proposed measures become public policy.

Conclusion

While caution, particularly with elective interventions like 'smart drugs' is always sensible, it is worth remembering that benefits unnecessarily forgone also have their price. This is one powerful reason why the law should be extremely reluctant to intervene paternalistically in the area of

CCEs. Many societies, including the UK, have for various reasons made dramatic distinctions between the control of drugs and a vast range of other enhancement modalities. There are understandable historical reasons for this, as well as less understandable prejudices and presumptions in play. In this essay we have argued for a more balanced and rational approach to access to CCEs, and, in particular, that the criminal law is an inappropriate mechanism to use to stand between citizens and their tenacious desire for self-improvement.

CCEs provide an opportunity to help us make the most of our skills and time, to become better individuals and, ultimately, part of a better society. Contrary to creating the much-dreaded 'sleepless society', these drugs will enable us to better meet the demands of the high-pressured, twenty-four-hour work-day we have already created for ourselves. It is important that realistic and proportionate limits are set to the amount of research deemed necessary before policy is determined. For, just as justice delayed is justice denied, so benefit delayed is benefit denied.

11 Dignity as a socially constructed value

Stephen W. Smith

Introduction

Human dignity[1] and its protection are important to legality, for intuitive reasons. It sounds beneficial to have dignity, and objecting to the possession of dignity in others is unlikely to gain significant support. It therefore provides a particularly strong debating device. Its importance is bolstered by its adoption in human rights treaties and national constitutions.[2] Unfortunately, dignity's intuitive appeal has often led to its not being discussed in detail. As a result, it becomes a concept that can be used in any and all situations. To take a few examples, those who wish to protect embryos from being destroyed for scientific knowledge invoke dignity.[3] Those hostile to the creation of clones do the same.[4] There are many other, at times idiosyncratic, examples.[5] Against this background the concept often remains ambiguous, and provides no real justification for a particular position. It becomes all things to all people. Indeed, some have questioned whether there is such a thing as human dignity or

[1] There are at least three conceptions of dignity that are used in legal and philosophical thought. The first conception of dignity relies on dignity as an aspirational characteristic, such as when we talk of people acting in a 'dignified' manner. The second conception is based on dignity as a characteristic of a situation with at least some external influences. This is the type of dignity used when we talk about 'death with dignity'. The final conception, which we shall focus on, is the concept of human dignity.

[2] D. Feldman, 'Human Dignity as a Legal Value: Part One' [1999] *Public Law*, 682. See also the Preambles to the UN Charter (1945), the Universal Declaration of Human Rights (1948), the International Covenant on Economic, Social and Cultural Rights (1966) and the International Covenant on Civil and Political Rights (1966). National constitutions that reference dignity include the German Constitution (Article 1), the South African Constitution (Article 1(a)) and the Belgian Constitution (Article 23).

[3] D. P. Sulmasy, O. F. M., 'Dignity and Bioethics: History, Theory and Selected Applications' in E. D. Pellegrino, A. Schulman and T. W. Merrill (eds.), *Human Dignity and Bioethics* (University of Notre Dame Press, 2009), pp. 469, 490–3.

[4] Ibid., pp. 493–6.

[5] Conseil d'Etat (27 October 1995) req. nos. 136–727 (Commune de Morsang-sur-Orge) and 143–578 (Ville d'Aix-en-Provence), cited in D. Beyleveld and R. Brownsword, *Human Dignity in Bioethics and Biolaw* (Oxford University Press, 2001), pp. 26–7.

whether it is merely another way of positing other ethical principles, such as a respect for persons or autonomy.[6]

While claims of dignity frequently arise in human rights contexts, it is also important in criminal law. The standard Millian analysis of state interference argues that the only ground for criminalisation that may interfere with the voluntary actions of individuals is harm.[7] Mill's primary concern was harm to others, as he saw this as the only justifiable reason for criminalisation.[8] Mill, though, never explains what constitutes harm. Fortunately, Joel Feinberg moves us forward, explaining that things can be considered harms if and only if they set back important interests that an individual has.[9] These include physical harms, but may also include some non-physical harms.[10] Interference with dignity, then, may provide us with arguments in favour of criminalisation of conduct.

In some cases, of course, violations of the dignity of others only provide additional reasons to claim a setback of interests sufficient to justify criminalisation. Non-consensual scientific investigation would be an example, but cases such as this are unlikely to cause much controversy probably because the claims for regulation do not depend upon the protection of dignity for their validity. More important for our purposes are two classes of claims which rely on dignity as providing the harm claim to explain the criminalisation of conduct. In these cases, the violation of dignity provides the sole reason for claiming harm. The first class of cases are those which involve persons who can have interests but it is only their dignity which has been infringed. Consider, for example, the famous French 'dwarf-tossing' case discussed by Beyleveld and Brownsword.[11] In this case, bars and clubs in France had provided the opportunity for individuals to participate in the 'sport' of 'dwarf-tossing'. This involved participants throwing another, willing participant down a padded alley.[12] These events were shut down by the French police on the basis that the activities in question violated the dignity of those being thrown.[13] This, despite the fact that all the participants were willing, and one of the dwarves argued against the ban on the basis that there were no setbacks to his interests involved in his

6 R. Macklin, 'Dignity Is a Useless Concept' (2003) 327 *British Medical Journal*, 1419.

7 J. S. Mill, *On Liberty* (Oxford University Press, 1991 [1859]), p. 14. 8 Ibid.

9 J. Feinberg, *Harm to Others: The Moral Limits of the Criminal Law* (New York: Oxford University Press, 1984), vol. I.

10 Ibid. For Feinberg, while a setback to interests is a necessary condition for something to be harmful, the setback to interests must also be a 'wrongful' one for criminalisation to occur (ibid., p. 34). The question of 'wrongfulness' is one that falls outside the purview of this essay, however.

11 Conseil d'Etat. 12 Ibid. 13 Ibid.

participation.[14] Indeed, he specified several benefits from his participation, including an income, a sense of self-worth and the ability to move in professional circles.[15] As there were safety procedures in place, none of the participants was physically harmed.[16] Hence, the only justification for criminalising the conduct is the protection of the dignity of the asserted victim (and that of others like him).

The second class of cases is those when the entity in question may not be capable of possessing interests at all. The most clear-cut cases are those related to the use of embryos. As far as we know, embryos do not have interests that can be set back. They do not feel pain. Nor is there any other likely source of harm sufficient to ground a claim for criminalisation. However, if they have dignity, that may be an interest sufficient to ground a claim of harm, and thus, possibly, of criminalisation of conduct relating to them. Other cases that would fall under this category include the treatment of those in persistent vegetative states (PVS), those with advanced dementia and those under the age of two, particularly newborns who have anencephaly (i.e. are missing at least part of their brain). These individuals may have more interests than embryos, but their interests are still often very restricted by their ability to create the sorts of interests that are commonly used to justify preventing harm.

These classes of cases illustrate a fundamental problem with using dignity in a criminal law context. Mill's harm principle is meant to limit the ability of states to interfere with conduct.[17] It is not meant to be a general grant of power. If the concept of dignity is not limited, it may become a 'catch-all' explanation of criminalisation, which is likely to metastasise: anyone can make a reasonable claim that something which is distasteful to them violates their dignity.

Equally, this is not a problem with the idea of human dignity per se, but a statement that unless we have an understanding of it, we cannot defensibly contain the concept. This requires answers to a number of questions. Which entities have dignity? Why do those entities have dignity? Furthermore, we must be able to explain what the relevant characteristics of dignity are, as well as what dignity entails. We must comprehend the ways in which dignity overrides, modifies and interacts with other relevant values we might encounter. We must know if it is possible to lose dignity and the relevant situations in which that might occur.

A full analysis of the concept of dignity is beyond the scope of this essay. Instead, it will explore the more limited questions of who has dignity and

[14] Ibid. [15] Ibid., p. 26.
[16] Feldman, 'Human Dignity', 702. [17] Mill, *On Liberty*, p. 14.

why those individuals possess it. For our purposes, it will be presumed that dignity is something that exists and can be used to legislate conduct.

An examination of who is covered by the concept of human dignity should start with the term itself. The term 'human dignity' implies, definitionally, that it must apply only to humans. The word 'human' in human dignity does relevant work in the concept. This does not mean that non-humans could not possess dignity. As a definitional matter, though, they do not possess human dignity. Furthermore, we can expand on this by looking at national constitutions and treaties. These, as a general rule, specify that dignity applies to all human beings. These documents often mention the 'dignity of all human beings', or use similar language, and therefore require that, if dignity exists, it must apply equally to all things within the category of human being.[18] We can then safely conclude that human dignity must apply to all human beings and only to human beings.

Our second initial claim follows logically from the first. If all human beings possess human dignity, there must be a reason for that. Consequently, whatever is used as a foundational claim to ground human dignity must be something that is possessed by all humans. Otherwise, there would be no justificatory reason for ascribing human beings with the same value if they did not all possess the same originating quality. Again, since this applies only to human beings, it must further be something only possessed by human beings. If our validating reason for having dignity applies to other things that are not human, there would be no rationale for limiting dignity only to humans.

Foundations for human dignity

Taking these two claims as a starting point, we can explore the foundations of human dignity. The first task is to explore whether the commonly used justifications can provide a sufficient grounding for human dignity. Two of the primary ones rely on a conception of humanity that is either theological or based on Kantian rationality.

The first justification for human dignity comes from the argument that humans were created in the image of God.[19] Those who take this view

[18] See Sulmasy, 'Dignity and Bioethics'.

[19] See, e.g., A. Schulman, 'Bioethics and the Question of Human Dignity', in Pellegrino *et al.*, *Human Dignity and Bioethics*, pp. 3, 8; R. P. Kraynak, 'Human Dignity and the Mystery of the Human Soul', in Pellegrino *et al.*, *Human Dignity and Bioethics*, pp. 61, 79. See also Pope John Paul II, *Evangelium Vitae* (25 March 1985) (www.vatican.va/holy_father/john_paul_ii/encyclicals/documents/hf_jp-ii_enc_25031995_evangelium-vitae_en.html).

do not necessarily agree on what being created 'in God's image' means. Still, whatever the 'image of God' means, all human beings possess it, and, as a result, we all possess human dignity.[20] While this argument could conceivably work for any theological position that corresponds to a creator-God, it is most commonly associated with the Judeo-Christian world view. It is also, unfortunately, extremely difficult to prove any of the argument's underlying assumptions. It requires us to accept that there is a God, capable of creating human beings in his/her image, and who indeed did so.[21] It is also necessary to accept that such an image of God is worthy of respect. There may be any number of additional assumptions created by the argument, although these are the most fundamental.[22] Perhaps none, all or some of the assumptions are true. Whatever the answer is to those questions, they are extremely difficult to prove. That, unfortunately for those arguing from the position of a 'God's image' foundation for human dignity, is a considerable problem. If you cannot provide support for some of the fundamental bases of an argument, you are unlikely to win over those not already predisposed to agree.[23] Those who are not will require proof.[24] This is particularly important, as a large percentage of the world's population does not accept the Judeo-Christian God. For those who do not, another justification for the dignity of all human beings is important. Consequently, the 'image of God' argument does not provide a satisfactory foundation for human dignity.

In part as a result of this, others have focused on Kantian rationality for that basis. Kant famously argued that human beings should be treated as ends in themselves.[25] What this means is that humans, because they are rational beings, are capable of fashioning their own

[20] Kraynak, 'Human Dignity and the Mystery', pp. 61, 79.

[21] It is worth highlighting that one does not necessarily need to accept that humans were created by God, even if one accepts that God exists and is capable of doing so. The only way that human existence provides a necessary proof of the existence of God is if God is the only thing capable of creating human life. That, however, is not a necessary step in the argument for human dignity. All that is required here is that God is capable of doing so and has done so, even if humans could have been created by other means.

[22] For example, one could consider the statement in n. 21 that God is the only being capable of creating the human species as a part of the argument. One could also consider the requirement that 'showing respect' as we conceive it is the best way to respect the fact that humans are created in God's image, but this is not as fundamental as those mentioned above.

[23] For an example of someone who does not accept the theological argument for dignity, see D. C. Dennett, 'Commentary on Kraynak', in Pellegrino *et al.*, *Human Dignity and Bioethics*, p. 83.

[24] Ibid.; Beyleveld and Brownsword, *Human Dignity*, p. 10.

[25] I. Kant, *Fundamental Principles of the Metaphysics of Morals*, trans. T. K. Abbot (Amherst, NY: Prometheus Books, 1988), p. 58.

ends in the world.[26] Since they can do so, they cannot be treated merely as means, but must be treated as ends in themselves.[27] Human dignity derives from this.[28] The importance of this is the hinging of the argument on this idea of rationality – which means being able to act from reasons. If humans cannot act from reasons, then they cannot possess rationality. If they are not rational, then they are not ends in themselves and not, presumably, deserving of dignity.[29] Unfortunately, not all human beings are rational, and none are probably rational at all times. Kantian rationality, however, does not require that human beings use their rationality in order to be entitled to respect. Instead, all that is required for the purposes of human dignity is the radical capacity for rationality.[30] Radical capacity is a general philosophical concept that can be used to deal with deficiencies in capacity. A capacity is an ability to perform some action or skill, even if one is not currently exercising that ability. I may therefore have the capacity to speak Chinese, even if I am not currently doing so. A radical capacity, on the other hand, is when an individual has the ability to develop a capacity even if they do not currently possess it. So, I have the radical capacity to speak Chinese if I could learn to speak Chinese, even if I cannot currently do so. With regard to rationality, all that is required is the ability to develop it, not the ability to exercise it currently.[31]

The benefit to requiring radical capacity for rationality rather than the ability to exercise rationality is that it provides for a greater number of possible rational humans. We can argue that a number of human beings who may not currently have the ability to act rationally can develop that ability, and thus have 'rationality' for the purposes of determining dignity. This is consistent with Kant, who seems to argue at various points that the capacity for rationality is sufficient.[32] For example, children and newborns are not rational beings under our general tests of rationality. They do, however, have the capacity to develop rationality, and therefore

[26] Ibid., p. 62. [27] Ibid., p. 58. [28] Ibid., p. 65.

[29] Kant is not entirely clear on this point. One of the versions of the categorical imperative applies to all human beings. At other times he refers to rational beings and argues that rational beings could be a larger group than humans (ibid., p. 35). For further discussion, see A. W. Wood, *Kantian Ethics* (Cambridge University Press, 2008), pp. 95–101.

[30] For a discussion about radical capacities, albeit in a slightly different context, see J. Finnis, 'A Philosophical Case Against Euthanasia', in J. Keown (ed.), *Euthanasia Examined* (Cambridge University Press, 1995), pp. 30–3; J. Finnis, 'The Fragile Case for Euthanasia: A Reply to John Harris', in Keown, *Euthanasia Examined*, p. 47.

[31] P. Lee and R. P. George, 'The Nature and Basis of Human Dignity', in Pellegrino *et al.*, *Human Dignity and Bioethics*, pp. 409, 410.

[32] Wood, *Kantian Ethics*, pp. 99, 103.

they would be covered. Unfortunately, while it may bring in any number of individuals who would not ordinarily qualify, it still cannot bring in all human beings. For example, while children under two may be able to develop the capacity for rationality, those in a PVS or with advanced dementia probably cannot. Furthermore, some newborns are born with anencephaly. In none of these three cases is it possible for someone to develop the capacity for rational thought, at least as far as we know. In the first and last cases, the individual in question is missing the relevant portion of the brain and thus the development of rational thought is impossible. Consequently, these individuals do not have the radical capacity for rationality any more than they have the actual capacity.

These individuals, then, would seem to lack human dignity, but this is inconsistent with the initial point. Human dignity, according to human rights documents and treaties, applies to all human beings.[33] If not every human being is rational, this cannot provide a sufficient grounding for a claim that all humans possess dignity. It cannot do this because it would not be able to support the claim that all humans are different from other things such as animals. If human beings are not different from other things, then it is not possible to claim that they ought to be subject to a special status. Rationality, then, can only provide a sufficient philosophical grounding if it is capable of applying to every human being.

We could attempt to argue that all human beings necessarily have the radical capacity for rationality as it is an essential part of being human.[34] Thus, even those who do not appear to have a radical capacity for rational thought must possess it. It is hard to see how this claim could be persuasive. Human being is a biological category. It therefore requires only biological methods for determining who is within a particular category. In other words, the defining characteristic of a human being is not the radical capacity for rational thought; it is that a particular entity has specific genetic material in a specific pattern. That is all that is required to be a member of the species homo sapiens. Claiming that rational thought is a fundamental part of being a human being is simply

[33] It is worth highlighting that, at the moment, we are only considering those humans who are necessarily part of the discussion about human dignity. So, those in PVS or anencephalic infants are clearly considered to fall under these sorts of claims. Embryos and foetuses are more problematic, but, at this stage, we can avoid a full discussion of them.

[34] Lee and George, 'The Nature and Basis of Human Dignity'; Finnis, 'Fragile Case', pp. 47–50 (although Finnis is primarily dealing with the claims of personhood, his argument requires that all human qualities apply, irrespective of whether a particular human being actually possesses them or even the capacity to possess them).

false. As a result, it is not possible to bootstrap these difficult individuals into the radical capacity for rational thought merely on the basis of their humanity.

One way round this problem that is more philosophically valid would be simply to deny that the relevant individuals in question are human. Thus, if rationality is really an essential component of being a human being and these individuals do not have the radical capacity for rationality, they simply cease to be human.[35] That argument flows much more logically from any initial claim that rationality is the ground for humans. However, this also seems to run counter to empirical facts we know about the world. Even if an individual in a PVS or with advanced dementia is not capable of rational thought, they are still human because they meet the biological criteria. While we cannot use the category of human beings to bolster claims about rationality, neither can we use rationality to deny something the characteristic of being human. Human being and rationality are not necessary characteristics of each other. Not all humans need be rational; not all rational beings need be human.

More fundamentally, one of the prime reasons for arguments for dignity is to protect those individuals who may lack rationality. If we return to the two ways in which the protection of dignity provided us with sufficient evidence of harm, the second class of cases were those in which the individual was incapable of having other interests that could be protected. The extent to which these individuals have interests is unclear, but, at best, their potential interests are likely to be very restricted. Consequently, there are only a limited number of ways these individuals can be protected. For example, they cannot rely on the autonomy claims used by others to protect their interests. Dignity, at least as a conception of interests that attaches to all human beings regardless of characteristics, provides a way to protect them without having to rely on the ability of the individual to make a specific claim for protection. Those most affected by dignity claims are likely to be those who cannot exercise rationality. As a result, we need to consider ways in which those individuals can be central to our conception of human dignity. Using rationality, however, means that those who should

[35] This sort of argument is not new. For example, when considering whether entities are entitled to moral status, some ground these status claims on whether the entity in question has particular characteristics, such as personhood. Those who do not are simply excluded from the class, even if they are human. See, further, on personhood, J. Glover, *Causing Death and Saving Lives* (London: Penguin Books, 1977); J. Harris, *The Value of Life* (London: Routledge, 1985); H. Kuhse and P. Singer, *Should the Baby Live?* (Oxford University Press, 1985); and M. Tooley, *Abortion and Infanticide* (Oxford: Clarendon Press, 1983).

be most protected are relegated to a subsidiary position. Indeed, considering that these individuals do not possess the radical capacity for rationality, they are most likely excluded.

A possible way out of this particular ethical quandary is to apply a precautionary principle in an effort to bring in all of those human beings still lacking the relevant criteria. A precautionary principle is a way of addressing the problem of the uncertainty of other minds.[36] Essentially, it is impossible to know for certain anything about other minds. Those entities we think are rational might not be; those things we think are not may be. We can never be completely sure. The ramifications of treating a rational being as non-rational are serious because we violate their interests if we classify them incorrectly. The method of dealing with this suggested by Beyleveld and Brownsword under the precautionary principle is to treat all other entities as rational beings as far as it is possible to do so.[37] This means that we should treat all things as capable of rational thought, as long as this does not adversely impact our ability to treat entities that are more likely to be capable of rational thought as rational beings.[38] In other words, our general aim is to engage in a probability exercise. If we are faced with two different beings and we cannot treat both as rational beings, then we should err on the side of the one that has a higher probability of being rational. Again, while this might expand the number of possible rational beings, it still seems unlikely that the possibility that someone like Anthony Bland (whose brain had devolved to a 'mass of watery fluid'[39]) has a greater capacity for rational thought than other entities, such as certain animals or plants.

All of this creates a significant problem. At the beginning, it was presumed that dignity is a valuable and useful ethical principle. However, no justification for its use as an ethical principle has been successful in covering all humans. In the next section, we shall consider an alternative.

Dignity as a socially constructed value

If it is impossible to create a foundation for dignity as an inherent value in human beings, it may be that the search for an inherent value is the

36 Beyleveld and Brownsword, *Human Dignity*, pp. 113–34. 37 Ibid., pp. 122–3.

38 Ibid. For a discussion of the use of precautionary principles in other settings, see S.W. Smith, 'Dignity: The Difference Between Abortion and Neonaticide for Severe Disability', in C. Erin and S. Ost (eds.), *The Criminal Justice System and Health Care* (Oxford University Press, 2007), p. 175, and S. W. Smith, 'Precautionary Reasoning in Determining Moral Worth', in M. Freeman (ed.), *Law and Bioethics, Current Legal Issues* (Oxford University Press, 2008), vol. XI, p. 197.

39 *Airedale NHS Trust* v. *Bland*, [1993] 1 All ER 821, 860 (*per* Lord Keith).

problem. Other than biological characteristics, nothing is inherent in all human beings. There is nothing we all share that can ground a claim for dignity. Fortunately, it is not necessary for any foundational criteria for dignity to be inherent, as that is not usually the primary concern. Dignity claims most often arise in respect of the welfare of those incapable of protecting themselves. Indeed, as shown earlier, most dignity claims are couched in terms of protecting human dignity, rather than fostering or developing it.[40] We primarily make claims in cases where people are being protected from being exploited or being forced into situations we consider dehumanising. Even when dignity claims are used in the context of research (such as the creation of embryos for stem cell experiments), it is used to argue for the protection of others – either the embryo as an entity in itself or future persons. As noted above, it is not about the development or fostering of capabilities. Instead, dignity claims provide an ethical brake or a reason to refrain from doing certain things. What is important is that dignity cannot be removed by others and is applied equally to those to whom we think it ought to apply. An inherent characteristic is one way we may address those concerns, but it is not the only way. An alternative is to have an inalienable concept. An inalienable concept is one that cannot be taken away from someone or something possessing it.

Considering dignity as an inalienable rather than an inherent part of being human still requires a foundation. If there is nothing intrinsic in human beings that provides such a claim, it will be necessary to look elsewhere. One important aspect of humanity that has not been explored in this context is the social one. Human beings are, in general, social beings. This social element plays an important role in our claims about dignity. When we use dignity claims, most of them relate to situations where an individual is being treated as a thing instead of as a member of that community. For example, in the French 'dwarf-tossing' case,[41] police stopped dwarf-tossing on the basis that it violated the dignity of the individuals in question, especially those who were being thrown by others.[42] In other words, the French police objected on the basis that the human projectiles were being used as things and not as people. This is despite the fact that all the participants were consenting.[43]

This idea of treating people as objects being unacceptable can be linked with previously explored notions of Kantian rationality, but it need not be. One can also argue that it is the social context of humanity that is at issue. What we are concerned about is how people are

[40] For a list of such documents see n. 2, above. [41] Conseil d'Etat.
[42] Ibid. [43] Ibid.

perceived by others and whether they are considered to be part of the social community.[44] In other words, the value of the participants in question was about whether they were considered part of a social group. The same can be said for dignity claims writ large. They are not claims about inherent characteristics, but whether individuals are being treated as part of a community.

As with any socially constructed value, it is important to consider which social group is at issue. Since the value here is human dignity, the individual in question needs merely to be part of the human social group – that is, a member of the species. Additional characteristics such as rationality are unnecessary. This is beneficial, as those individuals who might not satisfy those additional criteria are not required to, and these are often the people to whom dignity is most useful. It does not even matter that the individual is aware of their membership, merely that they be considered a member by those within the group (i.e. humankind).

To some, this may lessen the value of human dignity. Since it is something bestowed on an entity by others, it could be argued that it cannot be inalienable. If it has been given by others, perhaps they can take it away. We might consider this an especially difficult question if we have a particular group (for example, racist groups) that would routinely refuse the dignity of other groups and exclude them from the classification of those with dignity. It is worth emphasising that the group in question is humanity. Merely because a particular group wants to exclude others from that category does not mean that other human beings need to agree. We therefore need not fear that those in vulnerable groups will lose the protection of human dignity merely because there may be those within society who wish to exclude them. It is society as a whole that makes the determination.

What this highlights, however, is that any decision about whether a particular individual or class of individuals has human dignity is a conclusory statement. In other words, an argument is necessary for why a particular individual is part of a group. For a vast number of individuals, the argument will be simple. Being human is sufficient evidence of membership. More difficult problems may occur, though, when one considers other potential humans, such as those resulting from genetic engineering, embryos and cloned human beings. The unfortunate problem with this approach is that it is not easy to make a determination. The central question is whether these entities are sufficiently human-like as to be considered within the relevant category. In some cases, it seems

[44] For a similar argument, see D. Feldman, 'Human Dignity', 685–6.

unlikely that such a determination can be made until some of these entities are actually in existence. For example, we could consider a genetically engineered human–animal hybrid. Determining whether that hybrid is sufficiently human to qualify for human dignity is likely to be based on how much of the entity in question is 'human' and how much is 'animal'. It is impossible, however, to make such a determination in advance. This may provide little solace to those who think that entities such as genetically engineered human–animal hybrids should not be brought into existence in the first place. It is worth highlighting, though, that the conclusion reached here is merely a possible suggestion. It may be that society at large is able to make a determination in advance, based on what we believe to be the case. What is worth emphasising is that the important question is not what characteristics the potential entity might have, but whether or not we think it is sufficiently human to be part of the group. If it is not, then there is no reason why it need be granted human dignity from the start.

The uncertainty about the boundaries of the entity in question does not play a part, of course, in the matter of embryos. We are well aware of their nature. The larger question involved with embryos is that they seem to sit on the fault line between those things that are clearly human and those things that are not yet human (and are therefore not part of the group).[45] It is thus less a question of empirical proof about capabilities and more a question about the value that is placed on the embryo. It is a question of whether or not society believes that an embryo is close enough to those entities that are entitled to dignity to qualify for its protection. At the moment, there is no consensus about this. This controversy needs to be resolved before determining whether an embryo is entitled to dignity. It cannot, by itself, resolve that claim.

Conclusion

In conclusion, the best way to ascribe sufficient value to human dignity for it to be of use when determining harm is to base the concept on a socially constructed value as opposed to an inherent one. This provides a more solid grounding for a value that is to be ascribed to all human

[45] The words 'not yet' in this sentence are crucial. Human embryos are human to an extent, of course. The extent to which they are more human than, for example, sperm or egg cells, though, is a matter of controversy and the important dividing line. No one claims that sperm and egg cells, despite being human sperm and egg cells, are sufficiently human to qualify for human dignity. The question then becomes whether embryos are closer to sperm and egg cells or to those things we might more easily classify as having human dignity.

beings, as there is no characteristic that is possessed by all of us. The benefit of seeing dignity as a socially constructed value is that it allows us to focus the protection on those individuals who cannot rely on other legal and ethical principles for protection from the actions of others. If we accept that dignity applies even to those human beings who may not possess all of the mental or psychological characteristics of the rest of us, it can give us a strong justification to provide support and value to those individuals, despite the things they may lack. This means that, in relevant cases, it can provide us with a reason to criminalise conduct, but only after we have determined individuals to have value. Dignity cannot provide us with a reason to decide that a particular entity has value; it can only provide us with an interest to support once we have decided that value already exists.

This does not mean, obviously, that dignity ceases to be difficult to apply. There are many additional questions that need to be resolved. It is still not clear exactly what dignity requires of us. For example, we do not know yet whether human dignity requires that all human beings have a right to life. The exact remit and limitations of dignity still need to be examined. That, however, must begin with a clearer understanding of where dignity comes from. The purpose of this essay was to provide that understanding in a way consistent with its use in human rights law. From that foundation, we can begin to have a more thorough discussion about using dignity within the confines of the criminal law.

Part IV

Bioethics and criminal law in the dock

12 Can English law accommodate moral controversy in medicine? Lessons from abortion

Margaret Brazier

Introduction

In its recent report, *Animals Containing Human Material*, the Academy of Medical Sciences recommended that for the present a ban should be imposed on the transplantation of 'sufficient human-derived neural cells into a NHP [non-human primate] as to make it possible ... that there could be substantial functional modification of the NHP brain, such as to engender "human-like behaviour"'.[1] The Academy found itself addressing a very modern dilemma touching on morals and medicine. But the possibility of enhanced primates is just the latest conundrum needing the attention of bioethicists and lawyers, for a multitude of potential problems arise when law and biomedical science collide. As Mason and Laurie say, 'Medical law is catalysed by moral issues'.[2] Nor is it only modern science that provokes moral dilemmas. Nature may sometimes be at least partially to blame, as was seen in the litigation about whether or not to separate the Manchester conjoined twins.[3] In that case, the Court of Appeal was anxious to reject any notion that they sat as 'a court of morals'.[4] But as Mason and Laurie noted, the courts were 'clearly uncomfortable about being embroiled in this tragic matter, but ultimately the judges had to reach a decision'.[5] And they sum up the crux of the problem for the law. Debates in the chambers of the philosophers wax and wane for decades; the law must provide an answer that may often be no more than the least bad answer. And that answer

[1] Academy of Medical Sciences, *Animals Containing Human Material* (London: Academy of Medical Sciences, 2011), para. 8.2.3.

[2] J. K. Mason and G. T. Laurie, *Mason and McCall Smith's Law and Medical Ethics* (8th edn) (Oxford University Press, 2010), p. 2.

[3] *Re A (children) (conjoined twins: surgical separation)* [2001] Fam 147.

[4] Ibid., at 150, *per* Ward LJ.

[5] J. K. Mason and G. T. Laurie, *Mason and McCall Smith's Law and Medical Ethics* (7th edn) (Oxford University Press, 2006), p. 2. This helpful point is omitted from the 8th edition.

will often lack one single coherent rationale, but draw its force from different and overlapping considerations. The law will rarely resolve moral controversy; rather, it may find ways to accommodate the spectrum of moral disputes that make up a medico-moral controversy.

Difficult moral issues are not confined to medicine. Moral questions are present in many of the disputes that judges are asked to determine, and many areas of life that the law seeks to regulate. Complex moral dilemmas are not the exclusive domain of medical law. Many of the moral questions that medicine and science provoke are especially tricky, for at least three reasons:

(1) First and self-evidently, many medico-moral dilemmas touch on life itself – who or what do we allow to be born? How and when may you die? So judges will often affirm, as Lord Goff did in *Bland*, that 'the fundamental principle is the principle of the sanctity of human life'.[6] Then, equally often, their Lordships will seek to wriggle out of the dilemma by finding that the particular case does not offend the principle of the sanctity of human life.[7] I shall argue later that the historical origins for or against the law's respect for *sanctity* of human life, at least in the context of abortion, are less clear-cut than some might like to believe. History is invoked to support their particular moral case by very different partisans in what are often unhelpfully called pro-life/pro-choice debates.

(2) Tragic cases such as the conjoined twins notwithstanding, the proliferation of moral dilemmas triggered by medicine often results in large part from rapid developments in biomedical science, if only in medicine's capacity to correct defects in nature or, as is the case with abortion, carry out safely an intervention once fraught with peril.[8] It has become commonplace today to criticise judges and legislators for being unable to keep up with the pace of scientific developments and/or just not understanding science. I shall suggest that the roots of a degree of alienation between law and science are much older than we usually acknowledge. As David Faigman notes in his fascinating book, *Legal Alchemy*, 'Lawyers and scientists speak and think in different languages.' They are too often 'from different worlds'.[9]

[6] *Airedale NHS Trust* v. *Bland* [1993] 1 All ER 821 at 866.

[7] See, for example, *Re A*, at 203–4, *per* Ward LJ.

[8] This is equally true of the 'conjoined twins' dilemma. Advances in surgery made the separation feasible to save the life of one of the two and thus prompted the moral controversy.

[9] D. L. Faigman, *Legal Alchemy: The Use and Misuse of Science in the Law* (New York: W. H. Freeman and Company, 1999), p. xi.

(3) When scientific developments move from the laboratory to the clinic and the bedside, the very presence of law comes into question. If I seek help to end my pain and die, or a couple wish to take advantage of pre-implantation genetic diagnosis (PIGD) to avoid having a second child with cystic fibrosis, whose business is it? I shall again seek to show that the limits of the private domain of the physician–patient relationship are not an exclusively modern dilemma.

In this essay, I focus principally on the first two of these issues, which I examine mainly through the prism of debates on abortion over the centuries. The rich history of abortion law long precedes debates on the Abortion Act 1967. Reflection on that history reminds us that the common law is a tradition in which pragmatism plays a major role.

I restrict myself to the criminal law. Criminalising certain kinds of possible medical interventions or scientific developments signals that whatever its motivation, society outlaws that class of conduct and (in theory) places anyone challenging the law at risk of his liberty. 'Ethical objection' will not constitute a defence. The criminal law necessarily finds itself more readily engaged in medico-moral battles than other moral quagmires, just because of its centrality in the sanctity of life debates. Lord Devlin's[10] view that the criminal law should reinforce a general common morality across the board may command less support today, and identifying common morality is even harder now than was the case in 1965. So, for example, much of what Devlin argued in relation to sexual morality would gain less credence today, and H. L. A. Hart, his great opponent, would seem to emerge as the more prescient scholar.[11] In medico-moral controversy, the difficulty is this. Espousing Hart, not Devlin, I may (I do) hold that while my personal values dictate certain kinds of behaviour largely framed by Christian moral values, absent harm to others, I cannot seek to require that the law impose those values on everyone else. But once we move to look at issues such as embryo research or abortion, the 'harm' principle of itself (however formulated) is of little help. The question becomes, what is the nature of the relevant *other*?[12] For if I believe the *other* (embryo or fetus) to be intrinsically the same as me, I should seek to protect her from harm. Just because she looks different no more justifies saying she is beyond laws that protect

[10] P. Devlin, *The Enforcement of Morals* (London: Oxford University Press, 1965).
[11] H. L. A. Hart, *Law, Liberty and Morality* (London: Oxford University Press, 1963).
[12] See S. McGuinness, 'The Construction of the Embryo and Implications for Law', in M. Quigley, S. Chan and J. Harris (eds.), *Stem Cells: New Frontiers in Science and Ethics* (Singapore: World Scientific Publishing, 2012), pp. 51–84.

me than would be the case if I contended that all people over six foot tall were monsters who should be 'eliminated'. Now any philosopher reading this essay will wish to jump from here into debates on personhood and moral status. I regret that I shall disappoint them. I want rather to divert into the debate on how far any articulation of sanctity of life (or its absence) has a place in law, and to show that maybe evading the issue of personhood is, if not ideal, a sensible and pragmatic response for the law. Elsewhere I have asked the question: is the law on abortion in England a 'moral mess'?[13] In this essay I ask if 'messiness' is necessarily to be deplored. I am emboldened in such an 'un-academic' suggestion by Ronald Dworkin, who contends that in the context of abortion, the battle of opinions has been 'so intractable, so stuck in venom and hate', just because we have become so obsessed with the question of whether the fetus is a person, a question on which no argument can be decisive and no compromise is acceptable to the leading proponents in the debate.[14]

The role of the criminal law

Andrew Ashworth begins his text *Principles of Criminal Law* by stating that the 'chief concern of the criminal law is to prohibit behaviour that represents a serious wrong against an individual against some fundamental social value or institution'.[15] He qualifies this claim by explaining that criminal law is not *only* concerned with such conduct, and notes that the boundaries of the criminal law are 'explicable largely as the results of exercise of political power at certain points in history'.[16] In the Preface to his controversial and flawed[17] (but nowadays somewhat neglected work) *The Sanctity of Life and the Criminal Law*, originally published in 1957, Glanville Williams developed the argument (inter alia) that 'much of the law of murder rests on pragmatic considerations'.[18] Were my

[13] M. Brazier and E. Cave, *Medicine, Patients and the Law* (5th edn) (London: Penguin, 2011), p. 415.

[14] R. Dworkin, *Life's Dominion: An Argument about Abortion and Euthanasia* (London: HarperCollins, 1993), p. 238.

[15] A. Ashworth, *Principles of Criminal Law* (6th edn) (Oxford University Press, 2009), p. 1.

[16] Ibid., p. 2.

[17] See J. Keown and D. Jones, 'Surveying the Foundations of Medical Law: A Reassessment of Glanville Williams' The Sanctity of Life and the Criminal Law' (2008) 16 *Medical Law Review*, 85. Keown and Jones demonstrate cogently Williams' misinterpretations of Christian theology and, in particular, his distortion of doctrines of the Roman Catholic Church.

[18] G. Williams, *The Sanctity of Life and the Criminal Law* (New York: Alfred A. Knopf, 1974), pp. ix–x.

neighbour permitted to murder me and the state offered neither me protection nor my family redress, we might quickly revert to the law of vendetta. Braziers would ride out to exact vengeance on my killers. But, Williams went on, certain forms of criminal prohibitions on the taking of life, such as the criminalisation of infanticide, abortion and suicide, are 'rather the expression of a philosophical attitude than the *outcome of social necessity*'.[19] So, for Williams, infanticide (at any rate of a severely disabled baby) resulted in no adverse impact on society (assuming presumably that the parents were agreed on the undesirability of the baby's survival). Self-killing or the killing of a fetus or a newborn did not, in Williams' analysis, engage risk to social order, and thus for Williams did not necessarily justify the invocation of the criminal law. The mere fact that all the 'victims' had human DNA was not enough to bring them within the umbrella of laws protecting human life. A central tenet of Williams' thesis was that the criminal law on these issues was unduly based on Christian religious beliefs, and, in particular, he excoriated the influence of the Roman Catholic Church.

Abortion: a muddled history

Certain US legal historians (notably Cyril Means[20]) sought to assert that the common law courts eschewed engagement with abortion or infanticide, leaving such matters to the ecclesiastical forums for similar reasons to Williams' exclusion of abortion from the 'law of murder'. Killing that might lead to revenge or affect the King's Peace was the business of the Crown. 'Theological' questions about the unborn, or just born, child were the business of the Church. Means essentially contended that until Lord Ellenborough's Act of 1803, English and American women enjoyed a common law freedom to abort, and that legislative prohibitions on abortion were driven by the need to protect the life and health of women. Means' historical analysis is vigorously challenged by Keown[21] and, in a tome extending to 1,283 pages, by Dellapenna.[22] Dellapenna attacks Means' work as 'neither objective nor

19 Ibid., p. x.

20 See C. C. Means, Jr., 'The Law of New York Concerning Abortion and the Status of the Foetus 1664–1968: A Case of the Cessation of Constitutionality (1968) 14 *New York Law Forum*, 411: C.C. Means, Jr., 'The Phoenix of Abortional Freedom: Is a Penumbral Right or Ninth Amendment Right About to Arise from the Nineteenth-Century Legislative Ashes of a Fourteenth-Century Common-Law Liberty? (1971) 17 *New York Law Forum*, 335.

21 J. Keown, *Abortion, Doctors and the Law* (Cambridge University Press, 1988).

22 J. W. Dellapena, *Dispelling the Myths of Abortion History* (Durham, NC: Carolina Academic Press, 2006).

accurate'.[23] He notes that Means' work was funded by the National Association for the Repeal of Abortion Laws and describes his two key papers as propounding 'myths'.[24] I suspect that neither side of the debate is wholly right and lack the space to do justice to the wealth of argument.[25] Means' claim that prior to 1803 the common law allowed women liberty to abort is wrong. Prosecutions were rare, but it seems clear that to destroy the child in the womb, at least after quickening, was a criminal offence.[26] But claims that the common law was exclusively or even primarily motivated by a judicial endorsement that the fetus demanded protection as an innocent human being, that fetal life was sacred or inviolable, do not stand up to examination either.

Scrutiny of case law before 1803 shows mixed motives behind common law condemnation of 'abortionists'. Self-same judgments are invoked to contrary effect by partisans in the debate, selecting different quotations, sometimes from the same judgment, to make their case. While there is little doubt that procuring the death of the infant in the womb was an offence at common law, whether brought about by the woman herself or by a third party, many prosecutions related to the consequent death of the woman – unlawful act manslaughter.[27] Abortion was a highly dangerous procedure and limitations on abortion might properly be seen as at least in part designed to protect the pregnant woman as well as the fetus. Where a woman herself was indicted, at least part of her offence is often stated to be her *suicidal* intent. In their language, judges sent mixed messages; abortion, said Lord Tenterden, is 'against the law of God, injurious to the individual and *prejudicial to the State*'.[28] Gavigan cites judicial language condemning abortion attempted by the pregnant woman not so much for the killing of the fetus itself, but as unwomanly, 'contrary to the natural tenderness of the female sex'.[29] A commentator writing in 1832 regarded abortion as likely to corrupt sexual morals, allowing sexual pleasures to be indulged in with impunity.[30] And let us note that states (such as France and Romania), which, in more modern times, have sought absolute

[23] Ibid., p. 14. [24] Ibid., p. 15.

[25] For an excellent and more dispassionate analysis of the history of abortion law in England, see S. Gavigan, 'The Criminal Sanction as it Relates to Human Reproduction: The Genesis of the Statutory Prohibition of Abortion' (1984) 5 *Journal of Legal History*, 20.

[26] Gavigan, 'The Criminal Sanction', p. 29, and Keown, *Abortion*, pp. 3–11.

[27] Hale CJ contended that this could actually constitute murder of the mother; Gavigan, 'The Criminal Sanction', p. 26.

[28] *R* v. *Russell* (1832) 1 Moo C.C. 356 at 365.

[29] See the opening address of counsel for the prosecution in the trial of Eleanor Beare in Gavigan, 'The Criminal Sanction', pp. 28–9.

[30] Gavigan, 'The Criminal Sanction', p. 29.

prohibitions on abortion, have often done so to combat falling birth rates, rather than to affirm the sanctity of fetal life.

Whatever the truth about sanctity of life, the common law and abortion, the protection of fetal life was never the only card on the table. Keown, a prominent advocate of traditional Catholic ethics and the 'inviolability of life', acknowledges the role played by medical practitioners in the enactment of legislation relating to abortion in the nineteenth century.[31] For the emergent profession in the nineteenth century, criminalising abortion was part of a political agenda designed to embed the supremacy of a newly united (and only recently respectable) profession,[32] and in particular to endorse the dominance of the obstetricians over the midwives. A century later, relaxing restrictions by taking control of therapeutic abortions might be seen as an equally political manoeuvre, driven by a changed agenda.[33]

In the history of abortion and the common law, perhaps just one fact is beyond dispute: whatever crime abortion was, it was not homicide, 'if it be dead born it is no murder, for *non constat*'.[34] Where a woman quick with child killed her baby in the womb, it was 'a great misprision but no felony'.[35] Abortion was never wholly unlawful as it was in some civil jurisdictions. Keown recounts the evidence that in the nineteenth century, medical men 'were quite prepared to sacrifice the fetus to protect the life of the mother'.[36] To do so prior to the advent of safe Caesarean section, doctors resorted to the grisly procedure of craniotomy[37] – closely resembling what is described in the USA today as partial birth abortion.

The Offences against the Person Act 1861 formed the culmination of a comprehensive statutory prohibition of abortion (unlawfully procuring a miscarriage), be the attempt complete or inchoate, undertaken by the woman herself or a third party. But as *R* v. *Bourne*[38] demonstrated, abortion to prevent risk to the life or physical or mental health of the mother was not 'unlawfully' procuring a miscarriage. *Bourne* cannot be explained away simply as the application of a general liberty to defend third parties. The unfortunate girl operated on by Aleck Bourne was not at risk to her life, but at risk of severe 'mental and nervous injury'. Such a risk to a born human being would be unlikely to justify the killing of the tormentor as a proportionate response. However, the common law in

31 Keown, *Abortion*, pp. 27–48.
32 See M. Brazier, 'The Age of Deference: A Historical Anomaly', in Michael Freeman (ed.), *Law and Bioethics* (Oxford University Press, 2008), p. 464.
33 Gavigan, 'The Criminal Sanction', pp. 36–7.
34 *R* v. *Webb* (1601) 75 English Reports 1075.
35 *Coke's Institutes* 3, c.7.
36 Keown, *Abortion*, p. 60.
37 Ibid., pp. 60–2.
38 *R* v. *Bourne* [1939] 1KB 687.

England conceptualised the 'not yet born', it was not a full rights-holder. Yet it/she/he still retained *a* claim to protection. The nature and strength of that claim in the judgment of Macnaghten J in *Bourne* is hard to assess. He recites the mantra that the law 'has always held human life to be sacred'.[39] Yet in exonerating Mr Bourne he speaks of the 'anguish' to a 'normal decent girl'.[40] Had *she* killed her child at birth, would that anguish of itself exonerate her? She would of course have been able, in practice, to rely on the extenuation offered by the Infanticide Act 1922, a statute based on a profound misunderstanding of medicine. But pleading to infanticide would do no more than reduce the charge to manslaughter and not make the killing of the child lawful.

Science and certainty

An examination of the history of abortion sees the common law struggling with uncertainty. There is a 'thing' that counts for something, but little consensus about why, or how, or when. So from the mists of history to today, law has turned to medical science to attempt to find that missing certainty. Consider the controversy on 'quickening' or 'quick with child'. While Dellapenna seeks to argue that the term simply meant a live embryo, regardless of the stage of gestation,[41] most commentators agree that the evidence suggests that until that stage of maternal perception of the child in the womb, or at least some period into pregnancy, abortion was a lesser crime. In the gradual introduction of legislation in the nineteenth century, medical science was invoked to make nonsense of a distinction based on such unscientific and subjective evidence.[42] The notion of 'animation', often translated by theologians into 'ensoulment', was seen as 'disproved'. The embryo from conception was 'living', in a biological sense. Thus science drove legislation to criminalise the interference with pregnancy at any point imaginable in the nineteenth century. In a very different context, in the late 1980s, (expert) opinion led Judge Young in *Davis* v. *Davis*[43] to declare that frozen embryos created by the Davises' gametes were 'human' and therefore 'children', accepting Mrs Davis' expert Dr LeJune's thesis of 'tiny persons' and rejecting any distinction between pre-embryos

[39] Ibid., at 621. [40] Ibid., at 620.
[41] Dellapenna, *Dispelling the Myths*, pp. 132–3, 198–200.
[42] See Keown, *Abortion*, pp. 31–3, and Gavigan, 'The Criminal Sanction', pp. 34–5.
[43] *Davis* v. *Davis*, 1989 WL 14045 at *3: note that Judge Young's classification of the frozen embryos as 'tiny persons' was firmly rejected by both the Court of Appeals (1990 WL 130807) and the Tennessee Supreme Court (1992) 842 SW 2d 588.

and embryos.[44] Faigman castigates Judge Young for categorising the question as a scientific and not a legal question.[45]

In *Davis* v. *Davis*, science was relied on to seek to push a 'pro-life' agenda. In the seminal judgment in the USA, *Roe* v. *Wade*,[46] medical science was invoked to promote a different agenda, in establishing a constitutional right to abortion in the first trimester. Faigman[47] quotes Justice Blackmun that 'medical fact' showed mortality in abortion in the first trimester to be less than mortality in childbirth, but subsequently the reverse was true, engaging the state's interest in maternal health at the end of the first trimester. And in the third trimester, 'viability' justified state interest in the fetus. Medicine seemed to offer a neat compromise more clearly delineating maternal/fetal claims than the English Abortion Act, with its overt 'surrender' to medicalisation and doctors as gatekeepers.[48]

But 'medical facts' rarely stand still, nor are they as clear as lawyers might like. Safer abortions have extended the period of weeks in which statistically termination is safer than delivering, to sixteen to seventeen weeks. But is such statistical evidence itself 'viable'? Safety in terms of the risk of childbirth is very different for the healthy woman of twenty-five compared to the exhausted mother of eight, aged forty-three. As for viability, science can tell us today that very few babies will now survive at twenty-two weeks. If ectogenesis (artificial womb technology) becomes a possibility, the notion of viability may fly out of the window.[49]

Debates in this country have equally been obsessed with scientific evidence about viability and fetal awareness.[50] Experts are called by the legislature to ascertain when extremely premature babies *could* survive; when a fetus *might* feel pain. The laywoman (and the unwary legislator) is bombarded by disputed 'facts'. The moral stance of the expert on the 'fact' in question arguably influences her interpretation of evidence. The tradition of the common law as contest permeates the legislature as much as the courtroom.[51] Mrs Davis, seeking to protect her 'children', instructed (in Jerome LeJune) an expert whose eminence in medical science matched his ardent pro-life views. The House of

[44] See Faigman, *Legal Alchemy*, pp. 39–45. [45] Ibid., p. 43.

[46] *Roe* v. *Wade*, 410 US 113, (1973). [47] Faigman, *Legal Alchemy*, p. 106.

[48] Though note that the same 'evidence' is in practice relied on by English doctors to grant abortion on request, at least in the first trimester; see Brazier and Cave, *Medicine*, pp. 390–1.

[49] See A. Alghrani and M. Brazier, 'What Is It? Whose Is It? Re-positioning the Fetus in the Context of Research' (2011) 70 *Cambridge Law Journal* 51.

[50] See Royal College of Obstetricians and Gynaecologists, *Fetal Awareness – Review of Research and Recommendations for Practice* (London: RCOG Press, 2010).

[51] Faigman, *Legal Alchemy*, pp. 122–52.

Commons Science and Technology Committee, when reporting on the inquiry into *scientific* developments relating to the Abortion Act 1967, expressed concerns that some of the witnesses advancing testimony that was supposed to be confined to scientific issues were unduly influenced by their own moral stance on fetal interests, and that not all the experts had a track record in relevant areas of science or clinical practice.[52]

Facts often prove less objective than might be hoped when the law seeks to find answers in science. And what purpose do 'facts' serve? Were we to concede that twenty-one weeks six days marks viability and/or that at twenty weeks fetuses feel pain, what does that tell the legislator or the judge? Is it legitimate to destroy a twenty-one week five-day fetus because it is 'alive' too soon – it will be next year before such a baby could survive? That seems nonsensical. We might agree that causing pain to any sentient being is wrong. Anaesthetising the fetus answers that wrong. Biological science gives us *information*. Lawmakers often have only a partial understanding of that information, fail to understand that science rarely offers immutable certainties, and assume an objectivity often lacking when science is invoked in support of opinion.

The private domain?

Let me come briefly to the one other point – the domain of privacy. I do not seek to explore the impact of a right to privacy or Article 8 of the European Convention on Human Rights,[53] but simply to make a practical point about how medicine and abortion have changed over the years. Consider again the evidence that craniotomy was practised apparently without any legal challenge until women stood a fighting chance of surviving a Caesarean section. Eminent doctors (such as Dr Blundell) argued its legitimacy, saying tersely, 'with the dogmas of the divine, it is not my province to interfere'.[54] There seems little evidence of much public outcry – no Victorian pro-life movement. But then the procedures took place in the home, an engagement between doctor, midwife and family. Similarly, at the other end of life, doctors 'eased the passing' well outside today's bright lights of the hospital ward. In 1896, Hawkins J condemned the suggestion that a doctor should go to the police if he discovered a patient had undergone an illegal abortion as a 'monstrous

[52] HC Science and Technology Committee, *Scientific Developments Relating to the Abortion Act 1967 Twelfth Report 2006–7* (London: TSO, 2007), pp. 5–6.

[53] See *A. B. and C.* v. *Ireland* [2010] ECHR 2032; S. McGuinness, 'A, B and C Leads to D for Delegation' (2011) 19 *Medical Law Review* 476.

[54] Keown, *Abortion*, p. 60.

cruelty' and violation of the patient's trust.[55] A Victorian judge thus seemed to condone an illegal act, or at least prevent the crime from coming to the light of day.

It is the increasing role of the public domain that throws today's judges (as much as doctors) further into the murky waters of medico-moral controversy. That move is exacerbated by the legislature's reluctance to get its own feet wet and by the reactive nature of much legislation. Acts of Parliament touching on medical science are often out of date as the Royal Assent is graciously given. So it is the judges who are asked: should nurses be allowed to play a substantial role in lawful abortions?[56] Is using PIGD to create saviour siblings lawful?[57] Can nutrition and hydration be withdrawn from patients in a PVS?[58] Academics like to cavil at the Lordships' unwillingness to enter moral debate, the judges' references to engage in imaginative statutory interpretation. We also like to go for the jugular when judges do attempt to engage in ethical analysis. Perhaps we need to reflect more fully about whether we truly wish our judges to make law in this arena.

Inconclusive conclusions

I remain far from any conclusions in my own mind. Four tentative thoughts engage me:

(1) In shaping the common law's approach to human life, medical power, politics and fashion have played a major role. By medical I do not mean scientific opinion or evidence, but the politics of the medical profession, so I might amend Ashworth to say that the criminal law as it impacts on medicine is often 'explicable by medical politics at certain points in history'.

(2) However, we cannot avoid engagement with the 'sanctity' debate and follow Williams down the path of 'social necessity'. A reverence for life at all its stages of development and decay has been and is a crucial factor framing the development of laws. Such reverence is not simply a hangover from the teachings of the Churches.

(3) That means divergent opinions on what such reverence entails must be accommodated within the law. And opinion in many cases is simply not susceptible to objective proof. When law looks to science

[55] *Kitson* v. *Playfair* (1896), *The Times*, 28 March.
[56] *Royal College of Nursing* v. *Department of Health and Social Security* [1981] A.C. 800, HL.
[57] *R (on the application of Quintavalle)* v. *Human Fertilisation and Embryology Authority* [2005] 2 All E.R. 555, HL.
[58] *Bland.*

to assist in such dilemmas, the limits of scientific evidence need to be better appreciated and we need better means of obtaining less partial evidence – using the word partial in both senses of the word.

(4) Lawyers and scientists (and ethicists) need to understand that immutable certainties have never been discerned and never will be. So when we wish to engage the criminal law, should we be wary and demand a high standard of proof that the conduct in issue is (a) beyond the bounds of what can be contemplated in conformity with those values central to the criminal law, and (b) is better controlled and regulated by the criminal law than other instruments of state and society?

Disgraceful, I hear both the proponents of sanctity of life in its most orthodox form and their deadliest opponents cry. That is a 'fudge' and we need the criminal law to address these questions as only the criminal law can declare the 'truth' and the values by which society should live. The symbolic role of the criminal law is one of the themes that draws together the diverse essays in this work. I suspect we shall have to revisit it more often than might be comfortable and accept that in a diverse society all the law can do is seek to accommodate diversity. Consequently, there is no 'right' answer to the sanctity of life conundrum, but rather a number of answers that reflect diverging values. The challenge for the law is to pay equal consideration to these values and, in so doing, resolve moral controversy in a way that best fits the society in which it exists.

13 The case for decriminalising abortion in Northern Ireland

Marie Fox

Introduction

In Great Britain abortion is differentiated from other medical procedures since it is explicitly cast as criminal by the 1861 Offences Against the Person Act, with the 1967 Abortion Act (as amended) merely offering defences to health care professionals who terminate pregnancy within the grounds stipulated by the legislation. The 1967 Act, however, has never applied to Northern Ireland, with the consequence that no statutory grounds for abortion exist here, so that termination of pregnancy is a criminal offence subject only to minor exceptions carved out by the common law. There is widespread consensus that reform of UK abortion law is desirable, and particularly pressing in Northern Ireland.

This essay seeks to engage with the politics of abortion reform in this jurisdiction, addressing the most effective ways to frame arguments about abortion governance. I suggest that arguments for reform of abortion law will have greater purchase if they recognise the particular localised context, taking account not only of the politico-religious factors that have shaped prevailing conservative attitudes, but also of the advances that have been made in combating discrimination and promoting equality in this post-conflict society. I suggest that this approach also allows abortion to be understood as a moral choice. More specifically, I advance two main arguments. First, the most prominent and influential reform strategy has been to argue that the criminal law fails on its own terms by failing to live up to its ideology of certainty, so that Northern Irish law exists in a 'twilight zone'.[1] While conceding the appeal of such a strategy, I suggest that it is problematic given that such legal certainty is

Many thanks to Wei Wei Cao, Sheelagh McGuinness and to the editors of this volume for their helpful comments on earlier drafts.

[1] S. Lee, 'Abortion Law in Northern Ireland: The Twilight Zone', Public Consultation Document for the Standing Advisory Commission on Human Rights (SACHR) 1993, repr. in A. Furedi (ed.), *The Abortion Law in Northern Ireland: Human Rights and Reproductive Choice* (London: Family Planning Association, 1995), pp. 16–26.

illusory, and that arguing for it is out of line with trends elsewhere in the common law world. Thus, as Kirsten Savell has recently argued, 'the force and reach of the criminal law is diminishing'.[2] It follows, I argue, that feminist arguments should instead contest the deployment of criminal law to regulate pregnancy, by highlighting the material realities of criminalising abortion, as opposed to seeking to ensure that criminal regulation lives up to its own ideology. Second, I caution that although reforming Northern Irish law has become part of a raft of measures that lawyers and activists throughout the UK have sought, such reforms may prove counterproductive in the Northern Irish context. Since feminist strategies that are productive in one jurisdiction will not necessarily translate successfully to other jurisdictions, legal reforms must be framed with historical and cultural contexts in mind. Reform packages mooted for other parts of the UK have tended to bypass the moral dimensions of the debate and to be couched in consumerist rhetoric about choice and services. Instead, I advocate framing pro-abortion arguments as a matter of discrimination and injustice – discourses that have a particular resonance and appeal in Northern Ireland. In my view, human rights arguments that aim to demonstrate that Northern Irish women are discriminated against in comparison with their counterparts elsewhere in the UK are significantly more likely to command popular support than consumerist rhetoric, and have been underutilised in the abortion context.

The consequences of criminalisation: abortion in Northern Ireland

The law regulating abortion in the Republic of Ireland has generated recurrent criticism of the State's restrictive stance.[3] By contrast, comparatively little attention has been paid to the similarly restrictive regime in Northern Ireland. Since the 1967 Abortion Act never applied here, this jurisdiction too continues to be governed by the criminal prohibition on procuring a miscarriage in sections 58–59 of the Offences Against the

[2] K. Savell, 'Turning Mothers into Bioethicists: Late Abortion and Disability', in B. Bennett, T. Carney and I. Karpin (eds.), *Brave New World of Health* (Sydney: The Federation Press, 2008), p. 94.

[3] I. Bacik, *Kicking and Screaming: Dragging Ireland into the 21st Century* (Dublin: O'Brien Press, 2004); R. Fletcher, 'Post-Colonial Fragments: Representations of Abortion in Irish Law and Politics' (2001) 28 *Journal of Law and Society*, 568; R. Fletcher, 'Reproducing Irishness: Race, Gender and Abortion Law' (2005) 17 *Canadian Journal of Women & Law*, 365; S. Mullally, 'Gendered Citizenship and Reproductive Autonomy in Ireland' (2005) 27 *Human Rights Quarterly*, 78.

Person Act. It is thus truly a 'Victorian law,'[4] statutorily amended only by the Criminal Justice (Northern Ireland) Act 1945[5] and subject to limited common law exceptions carved out in a series of cases in the mid-1990s. These rulings – all concerning pregnant women who lacked capacity by reason of age or disability – confirmed that the 1939 English Crown Court decision in *R* v. *Bourne*[6] applied as a precedent and sanctioned termination only where the pregnant woman would otherwise be at serious risk of becoming 'a physical or mental wreck'.[7] However, although the women concerned were granted abortions, the *Bourne* ruling was construed restrictively. Hence MacDermott LJ emphasised in *Re A.M.N.H.*[8] that the risk to the woman must be 'real and serious', while in *Re S.J.B.*[9] and *Re C.H.*[10] it was stipulated that the risk must also be 'long term' and 'probable'. The overall effect is therefore to make the legal position broadly comparable to the much-criticised restrictive abortion regime in the Republic of Ireland. Moreover, since the case law to date has been concerned only with the implications of continuing the pregnancy for the pregnant woman's life or health, there is no legal ground in Northern Ireland to perform abortion on the basis of foetal disability, although it is widely accepted that most terminations here are carried out for this reason.[11] While acknowledging that the judges who heard the Northern Irish cases were constrained both by the facts of the cases and their timing prior to the Human Rights Act 1998, Eileen Fegan and Rachel Rebouche observe that they:

> failed to elucidate standards clear enough to inform medical practice in a variety of situations. The terms 'serious' and 'long term' imply a narrow scope for abortion legality without illustrating the type of grounds these labels might describe or the threshold sufficient to establish this risk.[12]

[4] A. Simpson, 'Abortion in Northern Ireland: A Victorian Law', in Furedi, *The Abortion Law in Northern Ireland*, p. 6.

[5] Which is equivalent to the Infant Life (Preservation) Act 1929, which created the offence of child destruction.

[6] *R* v. *Bourne* [1939] 1 KB 687.

[7] For more detailed discussion of this case law, see S. Lee, 'An A to K to Z of Abortion Law in Northern Ireland: Abortion on Remand', in Furedi, 'The Abortion Law in Northern Ireland'; T. McGleenan, 'Bourne Again? Abortion Law in Northern Ireland after Re K and Re A' (1994) 45 *Northern Ireland Legal Quarterly*, 389; D. Capper, 'The Condition of Abortion Law in Northern Ireland' (2003) 54 *Northern Ireland Legal Quarterly*, 320; B. Hewson, 'The Law of Abortion in Northern Ireland' [2004] *Public Law*, 234.

[8] *Re A.M.N.H.* [1994] N.I.J.B. 1.

[9] *Re S.J.B.* (1995, unreported).

[10] *Re C.H.* (1995, unreported). See also *Re Y.H.* (2003, unreported) and Re *N.T.* (2003, unreported).

[11] Lee, 'Abortion Law in Northern Ireland', p. 23.

[12] E. V. Fegan and R. Rebouche, 'Northern Ireland's Abortion Law: The Morality of Science and the Censure of Agency' (2003) 11 *Feminist Legal Studies*, 221, 229.

This notion that governance of abortion in Northern Ireland has consistently failed to live up to the standards of certainty and predictability expected of the criminal law has prompted recurrent criticism of the law, even from those who do not advocate a more permissive regime.[13] Thus, as Simon Lee argued in 1993:

> The law on abortion in Northern Ireland is so uncertain that it violates the standards of international human rights law. It could not withstand a challenge before the European Court of Human Rights ... Clarity is the key word. Some might argue that the law should be more permissive of abortions while some might argue that it should be more restrictive but all should agree that the present law is unacceptably uncertain.[14]

Today, notwithstanding legal challenges and attempts at parliamentary reform, it remains unclear precisely when women may lawfully access abortions. A woman who fails to convince a doctor that she is as desperate as the common law demands will be denied the procedure. Either she is forced unwillingly into motherhood or becomes one of an estimated 2,000 women who must travel to other parts of the UK.[15] According to the latest Department of Health statistics, over 5,500 women from both Irish jurisdictions travelled to England in 2010 for an abortion, accounting for 67 per cent of all terminations carried out for all non-resident patients in England and Wales.[16] Abortion thus becomes a 'problem exported',[17] the costs of which are imposed on pregnant women. The economic burden is considerable, since a termination may cost approximately £2,000,[18] but this is compounded by the isolation, loneliness and stigma captured vividly in narratives of abortion-seeking women. For instance, consider Magdelane, who became pregnant as a seventeen-year-old A level student:

[13] Lee, 'Abortion Law in Northern Ireland'. [14] Ibid., 16.

[15] Simpson, 'Abortion in Northern Ireland', p. 10.

[16] Department of Health, *Abortion Statistics, England and Wales: 2010* (London: Department of Health, May 2011); Family Planning Association, 'Latest Abortion Statistics Reveal Plight of Northern Irish Women' (press release), 25 May 2011 (see www.fpa.org.uk/pressarea/pressreleases/2011/may/latest-abortion-statistics-reveal-plight-of-women-in-northern-ireland).

[17] Simpson, 'Abortion in Northern Ireland', p. 10.

[18] A. Simpson, Submission to the UN Convention on the Elimination of Discrimination Against Women, July 2008 (see www2.ohchr.org/english/bodies/cedaw/docs/ngos/FPA.pdf). A Northern Irish Family Planning Association (fpaNI) press release in 2009 estimated that collectively these private abortions cost in excess of £1 million per annum, and pointed out that as UK taxpayers Northern Irish women were already contributing to the provision of NHS-funded abortions in the rest of the UK (see www.fpa.org.uk/pressarea/pressreleases/2009/25november2009).

> The time it took us to make all the arrangements ... meant I would be 14 weeks pregnant when I travelled to England ... My boyfriend and I saved and borrowed money to be able to travel to London. We didn't tell anyone that I was pregnant, as it was such a taboo subject ... I made an excuse to my parents ... and my boyfriend and I set off to fly to London. We couldn't afford to eat anything when we were there. The abortion had taken every penny we had and we just hadn't enough money ... I was taken into a consultation room, where I spoke with two doctors, and we discussed the procedure. After this, I was sent to a fee room, where I had to pay for the procedure ... I had £600 in cash in my purse, and the nurse asked me for £740. I nearly passed out. I explained that I had seen the price was £600 on the internet. The doctor who examined me said I was 16 weeks pregnant, not 14 as I thought. This meant the price had gone up, but I couldn't afford to pay for it ... In the end I was told there was nothing they could do for me. I didn't have the additional money and so I had to go back to Northern Ireland without having the abortion. I was absolutely devastated. I remember walking out of the clinic and falling on the ground crying ... On returning to Northern Ireland, I realised I would have to tell my parents ... After a lot of shouting and crying, my parents disclosed to me that they had been in the same situation when they were younger ... So my mum agreed to take me to London for an abortion. My mum took out a bank loan to cover the costs. I was 20 weeks pregnant at the time of the abortion, and I think it cost around £900 to have it done.[19]

Significantly, these broader questions of women's health, socio-economic circumstances and human rights have not been addressed in the case law, which has focused on the narrower issue of when the mental state of the pregnant woman will bring her within the defence accorded to doctors by *Bourne*. Essentially, women with an unwanted pregnancy in Northern Ireland must present themselves as absolutely desperate to have any chance of accessing abortion services, and it is unclear how sympathetic courts would be to more commonplace narratives. The law permits no space for them to present their decision to terminate as either rational or moral. Nor is there any provision for information or counselling services to be made available to a woman who is undecided whether to continue with her pregnancy.

Furthermore, the health risks to Irish women of typically having abortions at a much later stage than their British counterparts are compounded if they are deterred from seeking back-up treatment for any complications. Women who lack the resources to travel may also resort to illegal abortions. Dr Colin Francome's research, conducted in 1994, suggested that 11 per cent of Northern Irish GPs were aware of

[19] Magdelane's story is one of a number of real-life accounts that have been published on the fpaNI website. For other stories, see www.fpa.org.uk/campaignsandadvocacy/reproductiverights/abortioninnorthernireland/viralvideos.

illegal abortions taking place,[20] while in 1995 Audrey Simpson noted that since 1967 there have been five known deaths in Northern Ireland as a result.[21] Increasingly, a new era of 'backstreet' abortions takes the form of purchasing abortifacient drugs over the Internet.[22] Yet little information exists on either backstreet or legal abortions obtained by Northern Irish women. While abortion statistics have been available since 1998,[23] they offer little information as to the circumstances in which abortion is performed in the jurisdiction,[24] which is scarcely surprising given that abortion for fetal abnormality falls outwith *Bourne*. Thus, as Lee suggests, were the authorities to reveal the circumstances in which abortions are carried out or refused, this would likely expose them to litigation.[25] Consequently, legal uncertainty impacts on health professionals as well as the woman with an unwanted pregnancy. The lack of clarity leaves doctors open to the risk of prosecution, and likely entails that they are unduly cautious about performing abortions for fear of prosecution and stigma.[26] Such caution will also preclude the establishment of a supportive relationship with women seeking abortion services. Additionally, it means that health professionals possess no statutory rights to conscientiously object to the procedure.[27]

Clearly, however, the most pressing injustice is suffered by abortion-seeking women, cast by law as 'textual outlaws', who, as Alison Young argues, are 'sacrificed in order to maintain a fragile community'.[28] She notes:

[20] C. Francome, 'Birth Control and GPs in Northern Ireland' (1994, unpublished), cited in A. Simpson, 'Abortion in Northern Ireland', in E. Lee (ed.), *Abortion Law and Politics Today* (Basingstoke: Macmillan 1998).

[21] Ibid., p. 13.

[22] 'Women "Using Web for Abortions"', BBC News, 11 July 2008 (see http://news.bbc.co.uk/1/hi/7500237.stm). The report noted that a study in the British Journal of Obstetrics and Gynaecology found that of 400 'customers' almost 11 per cent had needed surgical procedures after taking the medication.

[23] Approximately seventy abortions are performed annually in Northern Ireland (fpaNI and Health Promotion Agency, *Abortion Fact Sheet* (2008)).

[24] K. Brennan, 'The State of Abortion Law in Northern Ireland', in J. Schweppe (ed.), *The Unborn Child, Article 40.3.3 and Abortion in Ireland* (Dublin: Liffey Press, 2008) pp. 247–75.

[25] Lee, 'An A to K', p. 34.

[26] Significantly, in *Re K* [1993] N.I. 268, even when Sheil J ruled that a suicidal fourteen-year-old could lawfully have an abortion as her case came within the tenets of *Bourne*, no doctor was willing to perform the operation in Northern Ireland, and she had to travel to England to have the procedure.

[27] Many commentators have viewed conscientious objection as simply a further obstacle to women seeking abortions (M. Thomson, *Reproducing Narrative: Gender, Reproduction and Law* (Aldershot: Ashgate, 1998), pp. 105–6).

[28] A. Young, *Imagining Crime* (London: Sage, 1996), p. 10.

Such an outlaw can never belong to the community, in that the community's very existence is founded upon her prior and continuing symbolic exclusion from it.[29]

Ruth Fletcher's work on the South of Ireland has demonstrated how abortion prohibitions function to shore up a sense of national identity.[30] While acknowledging that this concept is inevitably more contested and fractured in the North, by outlawing the actions of women who seek terminations, the law also constructs abortion as alien to community norms in Northern Ireland. Indeed, as such norms become increasingly contested, it would seem that groups opposed to abortion cling to them and the law that reinforces them more tenaciously. Yet the hypocrisy of maintaining restrictive regimes on both sides of the border only by virtue of the safety valve afforded by legal abortion services being available in neighbouring jurisdictions has been repeatedly criticised.[31]

Can law offer solutions?

Given the consequences of this hypocrisy for Northern Irish women and health professionals, a key question for reformers is what strategy is most likely to deliver effective reform. Tony McGleenan reads the Northern Irish abortion cases of the 1990s as 'part of a wider trend in the resolution of medical dilemmas by recourse to the law'.[32] And certainly the claim that law is unacceptably ambiguous suggests that the solution lies within law itself. For many, the quickest fix is extension of the 1967 Abortion Act to Northern Ireland. This has been British Medical Association policy since the 1970s, and was reaffirmed by its Representative Body in 2003:

> [T]his Meeting deeply regrets that women in Northern Ireland are denied access to the same abortion services available to women in the rest of the UK and requests that the government increase their efforts to ensure this anomaly is rectified as soon as possible.[33]

29 Ibid., p. 11. 30 Fletcher, 'Reproducing Irishness'.

31 E.g. S. McAvoy, 'From Anti-Amendment Campaigns to Demanding Reproductive Justice: The Changing Landscape of Abortion Rights Activism in Ireland, 1983–2008', in Schweppe, *The Unborn Child*, p. 41.

32 McGleenan, 'Bourne Again?', p. 390. This also ties in with the view of other commentators who point to a shift in modes of governance from medicalisation to legalisation (see R. Dingwall and P. Hobson-West, 'Litigation and the Threat to Medicine', in D. Kelleher, J. Gabe and G. Williams (eds.), *Challenging Medicine* (London: Routledge, 2006) pp. 40–61; K. Veitch, 'Medical Law and the Power of Life and Death' (2006) 2 *International Journal of Law in Context*, 137.

33 British Medical Association, *Abortion Time Limits: An Ethics Briefing Paper* (London: BMA, 2005) p. 8 (see www.bma.org.uk/ethics/reproduction_genetics/AbortionTimeLimits.jsp).

Of course, the 1967 Act has been much criticised and Lee has argued that it would be preferable to craft a new law for Northern Ireland, since the '[t]he opportunity is there for a genuine effort to establish a clear legal framework'.[34] Regardless of the precise form statutory amendment might take, commentators, judges and some politicians have viewed statutory reform as the appropriate mechanism for injecting certainty into an ambiguous legal position. Thus, Karen Brennan has urged, 'legislate we must',[35] and in 2009 Girvan LJ noted that 'there may well be a compelling argument for statutory clarification'.[36]

Yet whatever the merits of this call for legislative reform, and I would suggest that it is only compelling if it is premised on decriminalisation, politically it is unlikely to succeed. Girvan LJ's remark came just after the UK government had eschewed the opportunity to extend the 1967 Act to Northern Ireland. In so doing it displayed the reluctance to legislate that Therese Murphy and I have criticised elsewhere.[37] Politicians are typically unenthusiastic about abortion reform. On this issue, votes can only be lost, and the intractable nature of the debate in Northern Ireland means there is even less political will to act. Hence, when the Department of Health signalled its intention to revisit the Human Fertilisation and Embryology Act 1990, both pro- and anti-choice campaigners seized on this as a rare opportunity to attempt abortion law reform, as the 1990 Act itself had significantly amended the Abortion Act 1967.[38] Drawing on a number of reports[39] that had concluded that reform was necessary for abortion law to conform to modern principles of health care law (particularly its espousal of patient autonomy), pro-abortion campaigners sought parliamentary votes on the necessity of signature by two doctors, the permissibility of suitably trained nurses carrying out abortions, and the extension of the range of locations where terminations could lawfully be carried out. In July 2008, Labour MP Diane

[34] Lee, 'Abortion Law in Northern Ireland', p. 25.

[35] Brennan, 'The State of Abortion Law', p. 271, drawing on MLA John O'Dowd's speech in the 2007 NI assembly debate on the DHSSPS draft guidance (Department of Health, Social Services and Public Safety, *Guidance on the Termination of Pregnancy: The Law and Clinical Practice in Northern Ireland* (Belfast: DHSSPS, March 2009)).

[36] *Family Planning Association of Northern Ireland* v. *Minister for Health Social Services and Public Safety* [2004] NICA 39, at para. 10. In neither case was it clear whether such calls for statutory reform would decriminalise abortion.

[37] M. Fox and T. Murphy, 'Irish Abortion: Seeking Refuge in a Jurisprudence of Doubt and Delegation' (1992) 14 *Journal of Law and Society*, 454.

[38] R. Lee and D. Morgan, *Human Fertilisation and Embryology: Regulating the Reproductive Revolution* (London: Blackstone Press, 2001), Chapter 3.

[39] British Medical Association, *Abortion Time Limits*; HC Science and Technology Committee, *Scientific Developments Relating to the Abortion Act 1967 Twelfth Report 2006–7* (London: TSO, 2007).

Abbott tabled a series of amendments along these lines to the Human Fertilisation and Embryology bill, one of which would have extended the 1967 Act to Northern Ireland. As Sally Sheldon points out, such proposals were underpinned by evidence that they 'could contribute to easing access to abortion services, enhancing patent choice, reducing waiting times and bringing down the gestational age at which abortions are performed'.[40] Controversially, the Labour Party blocked the amendments, sparking allegations that they acted in order to secure the support of Northern Ireland's Democratic Unionist Party for the government's plans to legislate to allow 42 days' detention without charge.[41] Although the existence of any such deal was denied,[42] the absence of a plausible alternative explanation and the persistence of the rumours highlight both the political opposition to abortion in Northern Ireland and the potential for the issue to become a political football.[43] A further question is whether legislative reform strategies actually deliver the certainty their supporters claim. Thus, Lee's assertion that the 1967 Act as amended 'leave[s] the English law crystal clear, whether one likes it or not' may be doubted in the light of conflicting interpretations of grounds 1(i)(a) and (d).[44]

These obstacles to legislative reform have prompted both pro- and anti-choice activists to resort to litigation on each side of the Irish Sea.[45] In May 2001, the Northern Irish Family Planning Association (fpaNI) sought a judicial review of the Department for Health, Social Services and Public Safety's (DHSSPS) failure to fulfil its statutory duty to

[40] S. Sheldon, 'A Missed Opportunity to Reform an Outdated Law' (2009) 4 *Clinical Ethics*, 3.

[41] P. Toynbee, 'Resist the Medievalists. Women's Right to Abortion is a Private Matter', *The Guardian*, 9 May 2008 (see www.guardian.co.uk/commentisfree/2008/may/09/health.health?INTCMP=SRCH); P. Toynbee, 'Labour's Stitch-Up Will Deny Women Fundamental Rights', *The Guardian*, 21 October 2008 (see www.guardian.co.uk/commentisfree/2008/oct/21/health-health?INTCMP=SRCH); 'Legalise Abortion in Northern Ireland', *The Guardian*, 22 October 2008 (see www.guardian.co.uk/commentisfree/2008/oct/19/legalise-abortion-northernireland-labour?INTCMP=SRCH); Letter, 'Abortion Rights in Northern Ireland', *The Times*, 22 October 2008 (see www.thetimes.co.uk/tto/opinion/letters/article2064575.ece).

[42] M. O'Hara, 'We're Talking About Women's Lives', *The Guardian*, 17 October 2008 (see www.guardian.co.uk/lifeandstyle/2008/oct/17/women-northernireland?INTCMP=SRCH).

[43] Brennan notes that 'on more than one occasion a guarantee has been given to the effect that Westminster will not extend the Abortion Act 1967 to Northern Ireland' (Brennan, 'The State of Abortion Law', p. 272).

[44] These are the so-called 'social' and 'foetal handicap' grounds, both of which generate considerable uncertainty (see J. McHale and M. Fox, *Healthcare Law: Cases and Materials* (London: Sweet & Maxwell, 2007) pp. 878–86).

[45] R. Brownsword, *Rights, Regulation, and the Technological Revolution* (Oxford University Press, 2008).

ensure that women in Northern Ireland could access reproductive health care services. At first instance, Kerr J dismissed the application, doubting that the law was unclear or that departmental guidance would deliver any greater certainty. However, the Court of Appeal found that health professionals *were* confused about the legal rules on abortion and that the DHSSPS had failed to secure the provision of integrated health and personal social services to women seeking lawful termination of pregnancies.[46] It ordered the DHSSPS to inquire into the adequacy of provision of abortion services in Northern Ireland and to issue appropriate guidance following consultation. As Brennan notes, the Court of Appeal ruling also served to clarify the legal position, although arguably by making it more restrictive, since:

> [I]t is now plain that abortions are only lawful where the continuance of the pregnancy threatens the woman's life or would adversely affect her health ... abortion for any other reason, such as that of foetal abnormality, is not permitted.[47]

Five years later, following the establishment of a working group that solicited input from health professionals, guidance was issued.[48] Politicians stressed that the guidelines did not liberalise the legal position. For instance, the Ulster Unionist Party's John McAlister claimed that they 'recognise and reaffirm the present legal position – that, except where it is required to preserve the life of the woman, abortion is illegal in Northern Ireland'.[49] Yet journalist Liam Clarke contended, 'Don't believe the political spin – abortion has just become a lot easier in Northern Ireland',[50] and in 2010 the Society for the Protection of Unborn Children (SPUC) sought a declaration that the DHSSPS's decision to publish guidance was unlawful, asserting that it misrepresented the legal position, particularly in its 'failure to acknowledge the presumptive illegality of abortion in Northern Ireland' or 'to properly recognise the rights of the unborn child'.[51] Girvan LJ held that, while the guidance had not misrepresented the law on termination of pregnancy,

46 *Family Planning Association of Northern Ireland* v. *Minister for Health Social Services and Public Safety* [2004] NICA 39.
47 Brennan, 'The State of Abortion Law', p. 268.
48 Department of Health, Social Services and Public Safety, *Guidance on the Termination of Pregnancy.*
49 G. Moriarty, 'Guidelines Clarify When Abortion Is Legal in North', *Irish Times*, 21 March 2009. Confusingly, the SDLP's Carmel Hanna was quoted in the same report as stating 'the absolute legal ban on abortion remains unchanged'.
50 L. Clarke, 'At Last the Legal Fog on Abortion Is Lifted', *Sunday Times*, 22 March 2009 (see www.thesundaytimes.co.uk/sto/news/world_news/article157140.ece).
51 *Society for the Protection of Unborn Children, Re Judicial Review* [2009] NIQB 92, para. 7.

it was misleading with regard to counselling and conscientious objection. The DHSSPS was ordered to withdraw the guidance and reconsider it in the light of this judgment.

The limitations of fpaNI strategy in terms of securing legal clarity are obvious. As Girvan LJ acknowledged:

> Where the existing law is unclear such advice as is given would have to be qualified and would have to be presented in such a way as to show that what the Guidance provides is advice which is subject to developing case law keeping in mind that in an area of law which is not wholly developed there may be incremental developments in the development of case law and the present law cannot be stated with complete certainty.[52]

What now?

Given that medico-legal commentators had identified lack of clarity as the central problem, it would appear that attempts at reform, both through the parliamentary process and litigation, have been a resounding failure. However, it is arguable that law is simply more indeterminate than such reformers allow and that the legal certainty they pursued is a chimera.[53] In any event, it would seem that the 2008 reforms to the Human Fertilisation and Embryology Act represented the final opportunity for Westminster to extend the 1967 Act to Northern Ireland. With the anticipated devolution of criminal justice matters to the Northern Ireland Assembly, 'the tenor and outcome of the previous assembly debates on abortion [mean] it is highly probable that if Northern Irish politicians legislate on this issue, it will be with the objective of ensuring that abortion is only available in very restricted circumstances'.[54] The danger, then, in pursuing legislative reform through the Assembly is that the criminalisation of abortion may be even further entrenched. Certainly, the scrutiny that any future DHSSPS guidance will attract suggests that it is likely to adhere very closely to the current restricted position, while relying on professional rather than statutory guidance also 'obviates the opportunity for accountable debate about the shape of abortion law'.[55]

Consequently, as in the Republic of Ireland, it seems that the most promising option for reformers is to pursue arguments grounded in how women are denied their rights to privacy, health and non-discrimination. Space precludes any attempt to speculate on the likelihood of such claims succeeding, although they have made only limited headway in

[52] Ibid., para. 10.
[53] A. Norrie, *Crime, Reason and History* (2nd edn) (Cambridge University Press, 2001), Chapter 1.
[54] Brennan, 'The State of Abortion Law', p. 273. [55] Lee, 'An A to K', p. 45.

contesting the Republic of Ireland's abortion law before the European Court of Human Rights,[56] and it is not clear that UK courts would be any more receptive to claims under the Human Rights Act 1998. Yet all is not bleak, since many valuable advances have been made as a result of pro-choice reform strategies. In the first place, these legal struggles have placed abortion reform on the political agenda, and have increased public awareness of the issue. Thus, in the run-up to the parliamentary debates in 2008, the issue attracted significant media coverage, much of it recounting women's narratives, and citing a poll that found that 62 per cent of the Northern Irish public supported abortion in cases of rape and incest.[57] And, as Fletcher observes of the fpaNI campaign:

> By airing women's abortion stories in court and in the media, the negative effects of abortion restrictions on women's lives have become more publicly obvious.[58]

Second, and perhaps most importantly,[59] the fpaNI case did go some way to position abortion as a legitimate medical treatment or service rather than a criminal action. Some feminist legal scholars have criticised the fpaNI tactics, arguing that a strategy grounded in women's health needs does not sufficiently challenge the cultural and legal norms of Northern Ireland.[60] Yet in many respects, this represents the main strength of the strategy, since there are compelling reasons to work with rather than against cultural norms. As Morgan Healey argues, drawing on the work of Ferree *et al.*, 'frames' or 'central organising ideas that provide coherence to a designated set of idea elements'[61] construct the parameters of debates – 'they set up what language and meaning is included based on social/cultural standards'.[62] In this vein, I would argue that the challenge for those who seek a more permissive abortion regime is to find

[56] See, most recently, *A, B & C* v. *Ireland* [2010] ECHR 2032; S. McGuinness, 'A, B and C Leads to D for Delegation!' (2011) 19 *Medical Law Review*, 476.

[57] Toynbee, 'Resist the Medievalists'; O'Hara, 'We're Talking About Women's Lives'; Editorial, 'Legalise Abortion in Northern Ireland', *The Observer*, 19 October 2008 (see www.guardian.co.uk/commentisfree/2008/oct/19/legalise-abortion-northernireland-labour?INTCMP=SRCH).

[58] R. Fletcher, 'Abortion Needs or Abortion Rights? Claiming State Accountability for Women's Reproductive Welfare' (2005) 13 *Feminist Legal Studies* 123, 133.

[59] Much to the ire of anti-choice groups (Precious Life, *Briefing Paper on the DHSSPS' Guidance on the Termination of Pregnancy: The Law and Clinical Practice in Northern Ireland* (Belfast: Precious Life, 2009).

[60] Fegan and Rebouche, 'Northern Ireland's Abortion Law', p. 232.

[61] M. M. Ferree, M. Gamson, J. Gerhards and D. Rucht, *Shaping Abortion Discourse: Democracy and the Public Sphere in Germany and the United States* (Cambridge University Press, 2000) p. 105.

[62] M. Healey, '"I Don't Want to Get into This, It's Too Controversial": How Irish Women Politicians Conceptualise the Abortion Debate', in Schweppe, *The Unborn Child*, pp. 65–85, at 66.

a frame other than criminal law that captures the narratives of abortion-seeking women in Northern Ireland.

Framing arguments for decriminalisation

The health frame adopted by the fpaNI is clearly one contender. It fits with global trends where 'the reach of the criminal law is dwindling in respect of ... clinical encounters, particularly where the doctors concerned are eminent and are acting within standards accepted as ethical by the medical profession'.[63] Such reasoning also underpinned the arguments of pro-choice campaigners in their attempts to hitch abortion reform to the 2008 Human Fertilisation and Embryology Bill. These were couched in terms of promoting women's health and drawing analogies with other medical treatments. Thus, Sheldon has advocated that owing to its normalisation as a routine medical procedure for women of child-bearing age, abortion should be regulated in the same way as other medical interventions.[64] Yet, while I agree with the fpaNI's strategy of positioning abortion as medical treatment,[65] I would demur from representing abortion as equivalent to other surgeries. By erasing the factors that make abortion a morally troubling decision for some women, this argument risks essentialising women's experience of abortion, and may prove counterproductive in a jurisdiction with socially conservative, albeit shifting, attitudes to abortion. It exemplifies Julia Hanigsberg's charge that feminist theorists have 'inadequately theorized the importance or meaning of "intrauterine life"'.[66] She contends that abortion debates that ignore the existence of the foetus 'allow ... only the "pro-life" side of the political coin to evince any concern for life'.[67] I agree with Hanigsberg that space must be found within feminist discourse on abortion to acknowledge that many women do not experience abortion as analogous to surgery to alleviate back pain.[68] Consequently, rather than

[63] Savell, 'Turning Mothers into Bioethicists', p. 109. Though note that the declining influence of criminal law exists alongside the greater reach of law generally (see n. 32, above).

[64] Sheldon, 'A Missed Opportunity', p. 3; E. Jackson, *Medical Law: Text, Cases and Materials* (Oxford University Press, 2010), p. 709.

[65] This strategy also fits with an emerging international human right to health, which is violated by restrictions on abortion (R. Silfris, 'Restrictive Regulation of Abortion and the Right to Health' (2010) 18 *Medical Law Review*, 185).

[66] J. Hanigsberg, 'Homologizing Pregnancy and Motherhood: A Consideration of Abortion' (1995–6) 94 *Michigan Law Review*, 371, 380.

[67] Ibid.

[68] C. Gilligan, *In a Different Voice* (Cambridge, MA: Harvard University Press, 1982), pp. 70–98; K. McDonnell, *Not an Easy Choice: A Feminist Re-Examines Abortion* (Toronto: The Women's Press, 1984).

seeking to depoliticise or demoralise abortion decision-making, I suggest that it is preferable to confront the morality of abortion decision-making head-on, and seek to present women's abortion decisions as moral choices, even though they entail the ending of life.[69] Drawing on real-life narratives[70] allows the moral nature and material context of women's abortion decisions to emerge.

Fegan and Rebouche contend that to effect the necessary shift in cultural ethos, a more radical approach that presents women's agency in reproductive decision-making as fundamental must be adopted.[71] Yet it is doubtful whether such arguments will be sympathetically received. As Murphy and I have argued in relation to abortion debates in the Republic, feminist theorising in Ireland remains in its infancy, so that strategies employed by feminist scholars in other jurisdictions may not work in Ireland, where the cultural heritage imposes constraints on which arguments are likely to succeed.[72] While debates have undoubtedly moved forward on both sides of the Irish border since 1992,[73] it remains crucially important how arguments in favour of abortion are framed. As we suggested, it is difficult to transpose solutions from one jurisdiction to another when 'arguments grounded in the language of reproductive choice and autonomy (which are not part of the daily currency of Irish women's conversation), may alienate those whose very identity may be grounded in their status as wives and mothers'.[74] Although Fegan and Rebouche allude to shifts in public attitudes, it is questionable that they amount to a 'sea change'.[75] As Fletcher notes, 'public endorsement of a pro-choice position is still rare',[76] and public bodies have been extremely wary of engaging with the issue. For instance, in a report advising the Secretary of State on a Bill of Rights

[69] M. Fox, '"A Woman's Right to Choose": A Feminist Critique', in J. Harris and S. Holm (eds.), *The Future of Human Reproduction: Choice and Regulation* (Oxford: Clarendon Press, 1998), p. 77.

[70] M. Boyle and J. McEvoy, 'Putting Abortion in its Social Context: Northern Irish Women's Experiences of Abortion in England' (1998) 2 *Health* 283; A. Rossiter and M. Sexton, *The Other Irish Journey* (London: Voice for Choice, 2001).

[71] Fegan and Rebouche, 'Northern Ireland's Abortion Law'.

[72] Fox and Murphy, 'Irish Abortion', p. 462.

[73] McAvoy, 'From Anti-Amendment Campaigns'.

[74] Fox and Murphy, 'Irish Abortion', p. 462.

[75] Though Fegan and Rebouche rightly observe that popular hostility to abortion in Northern Ireland has probably been overstated, citing a 1998 *Northern Irish Life and Times* survey, demonstrating support for broadening the grounds on which abortion is available, and a *Belfast Telegraph* poll in 2003, in which a majority of respondents supported legalising abortion on demand (Fegan and Rebouche, 'Northern Ireland's Abortion Law', p. 233).

[76] Fletcher, 'Abortion Needs', 124–5.

for Northern Ireland, the Northern Ireland Human Rights Commission (NIHRC) refused to include the issue of reproductive rights,[77] while similar reluctance to confront abortion rights has characterised the work of the Committee on the Administration of Justice – the leading independent human rights organisation. Therefore, while Sheldon rightly observes of Great Britain that 'abortion has become part of the mainstream, an accepted and necessary – if often painful – part of contemporary life',[78] this is simply not true of Northern Ireland. Consequently, it remains crucial, so far as possible, to work with rather than against cultural norms in seeking to change attitudes. My preferred approach would therefore focus on framing the criminalisation of abortion as a form of discrimination against Northern Irish women. This taps into earlier and powerful feminist theorising of abortion as gender discrimination. As Catherine MacKinnon argues, criminal abortion statutes:

> make women criminal for a medical procedure that only women need, or make others criminal for performing a procedure on women that only women need ... Because pregnancy can be experienced only by women, and because of the unequal social predicates and consequences pregnancy has for women, any forced pregnancy will always deprive and hurt one sex only as a member of her gender. Just as no man will ever become pregnant, no man will ever need an abortion, hence be in a position to be denied one by law.[79]

Non-discrimination and equality discourse is a frame that has become widely accepted in the employment sphere and also more broadly in Northern Ireland,[80] and, as Fletcher notes, 'in a political context which has seen conflict and suspension of the rule of law, appeal to the ideals of state fairness and justice has strong resonance'.[81] As long ago as 1987, such discourses were being applied to abortion. In that year, an international tribunal organised by the (now defunct) Northern Ireland Abortion Law Reform Association (NIALRA) found that requiring

[77] NIfpa responded with a press release attacking the NIHRC stance and noting that since the introduction of the Human Rights Act there had been no 'real effort to address the human rights implications of Northern Irish abortion law' – *FPA in Northern Ireland responds to the Human Rights Commission report omission of reproductive rights* – available at http://www.fpa.org.uk/pressarea/pressreleases/2008/humanrightscommissionreportomitsreproductiverights (accessed 18 March 2011).

[78] Sheldon, 'A Missed Opportunity', p. 4.

[79] C. MacKinnon, 'Reflections on Sex Equality Under Law' (1991) 100 *Yale Law Journal*, 1281, 1319–20.

[80] B. Hepple, 'Discrimination and Equality of Opportunity: Northern Irish Lessons' (1990) 10 *Oxford Journal of Legal Studies*, 408–21; C. Harvey (ed.), *Human Rights, Equality and Democratic Renewal in Northern Ireland* (Oxford: Hart, 2001); C. Harvey, 'Contextualised Equality and the Politics of Legal Mobilisation: Affirmative Action in Northern Ireland' (2012) 21 *Social and Legal Studies*, 1–28.

[81] Fletcher, 'Abortion Needs', p. 126.

Northern Irish women to pay for terminations in Britain discriminated against the most economically disadvantaged women. It noted:

> Women in Northern Ireland should have at least the same right as women in the rest of the UK. We believe the extension of the Act will reduce the unnecessary and inhumane suffering women in Northern Ireland are forced to endure through unwanted pregnancy.[82]

Similarly, a 1995 symposium organised by fpaNI concluded that 'the difference between abortion laws which apply to women in Northern Ireland and women in the rest of the UK amounts to discrimination against women in the Province [*sic*]'.[83] Besides its cultural relevance, a further advantage of framing abortion in terms of discrimination and violation of women's human rights is that it facilitates the casting of the argument in explicitly moral terms. Of course, there are risks in casting the debate in this way, since 'abortion in Northern Ireland has been the subject of moral debate, often in language of extremism and hysteria rather than an informed political discussion'.[84] Yet in order to challenge the claims of anti-choice groups to the moral high ground, I would argue that there are good reasons to present the abortion decision as a moral one that women can be trusted to make. To bypass ethical debate by positioning abortion as simply another consumer choice is symptomatic of what Jonathan Montgomery has termed the 'demoralisation' of health law.[85] In Northern Ireland I would urge that the most effective tactic is to challenge the construction of abortion-seeking women as outlaws (by both anti-choice groups and criminal law), by conveying the complex realities of how and why their decisions are reached and the consequences for women's health and lives of denying them the right to make abortion decisions. This would highlight how women in Northern Ireland are disadvantaged, both in comparison to Northern Irish men, who cannot become pregnant, and in comparison to women in the rest of the UK or Europe.

Conclusion

In this essay I have sought to argue that there is a moral as well as a health-related case to be made for why women terminate pregnancies, and that this makes it wrong to criminalise the actions of abortion-seeking women

82 NILRA, 8 October 1987.

83 M. Clark-Glass, 'Introduction', in Furedi, *The Abortion Law in Northern Ireland*, pp. 2–5.

84 Simpson, 'Abortion in Northern Ireland', in Furedi, *The Abortion Law in Northern Ireland*, p. 14.

85 J. Montgomery, 'Law and the Demoralisation of Medicine' (2006) 26 *Legal Studies*, 185.

(and those health professionals who wish to assist them). I have suggested that allowing the stories of women to be told is the most effective way to win the 'hearts and minds' battle that must be waged over abortion in Northern Ireland. I believe it is crucial not to cede the moral arguments to 'pro-life' groups, especially when the legal arguments are so hard to win, and when law cannot supply all the solutions. Thus, notwithstanding the power of the criminal law to negatively impact on women's lives, feminist legal scholars also need to be mindful of its limitations. While it continues to be important to work with health professionals to secure the provision of services that promote rather than damage women's health, I would argue that we also need to acknowledge that foetal life is part of the equation, making it counterproductive to deny that abortion differs from other medical procedures. However, by framing the criminalisation of abortion as a matter of discrimination and violation of the human rights of Northern Irish women, we can work within cultural norms to make a case for legitimising their actions, even at the cost of foetal life.

14 The impact of the loss of deference towards the medical profession

José Miola

Introduction

This essay seeks to examine the traditional deference shown by both the civil and criminal courts towards the medical profession when dealing with questions of medical accountability and areas of bioethical controversy, such as human tissue and organs. In the past, doctors in the UK, were, uniquely, effectively allowed to set their own legal standards. I argue here, however, that the current situation presents something of a mixed picture. On the one hand, a change in approach on the part of the civil law has had ramifications for the criminal law, as the legal test for gross negligence manslaughter is based on the civil law test. Furthermore, the Human Tissue Act 2004 perhaps represents a new approach by Parliament. However, in the criminal courts, some of the same assumptions that led to the deference in the civil law remain, and doctors are still in a privileged position. Thus, while the legal changes have had some effect, an underlying deference can still be found and can be seen to work to the benefit of doctors. The aim of this essay is to chart the reduction in deference in the civil law, and then to assess whether it has been replicated by the common law and Parliament in the criminal sphere.

The existence of deference in the civil courts

That the courts used to hold an overly deferential attitude to the medical profession is undeniable. Doctors were simply assumed to seek to act in the best interests of their patients at all times, and this assumption had dissuaded courts from using the criminal law rather than the civil law to right wrongs.[1] An example of this is the issue of medical treatment provided without consent, which may be actionable either as a criminal or civil battery, or as negligence. The courts have repeatedly voiced their

[1] See, for example, J. Harrington, 'Privileging the Medical Norm: Liberalism, Self-Determination and Refusal of Treatment' (1996) 16(3) *Legal Studies*, 348–52.

preference for the latter to be used, since battery, 'with its connotations of violence and criminality, is seen as an inappropriately stigmatic label for doctors seeking in good faith to act in their patients' interests'.[2] However, the use of negligence legitimised the almost wholesale abrogation of responsibility by the courts to the medical profession.

Nothing encapsulates this deference more than the *Bolam* test in the civil law, which is also relevant to the criminal law with respect to charges of gross negligence manslaughter.[3] For a charge of gross negligence manslaughter to be successful, the prosecution must first prove negligence in the civil sense,[4] then that it went so far beyond this that it would be more appropriate to punish the defendant through the criminal law rather than the civil law – in other words, that the negligence is 'gross'. This was explained in the case of *R* v. *Bateman*, where the court defined 'gross' negligence as occurring when:

> the facts must be such that ... the negligence ... went beyond a mere matter of compensation between subjects and showed such disregard for the life and safety of others to amount to a crime against the state and conduct deserving [of criminal] punishment.[5]

Perhaps more pithy is Lord Mackay's observation in *Adomako* that the conduct must be 'so bad in all circumstances as to amount to a criminal act or omission'.[6] Given the fact that a charge of gross negligence manslaughter cannot succeed without the civil wrong first being proved, it can be seen to be crucial to the construction of the offence. Moreover, analysis of the law, and how it has changed, provides a valuable insight into a potential change in judicial attitudes towards the medical professionals and the amount of deference that they are to be shown by judges.

The *Bolam* test, designed as defining the standard of care in a civil action for medical negligence, although later spreading throughout medical law, was originally interpreted as providing that so long as a medical practitioner could adduce honestly held evidence on her behalf that some other doctors might have done the same as her, she would have been acting reasonably and thus not below the standard of care required by law. What this meant in practice was that, so long as the defendant medical practitioner could find

2 Ibid., 352. For an example of the courts' preference for the use of negligence over trespass, see *Chatterton* v. *Gerson* [1981] QB 432, where trespass was defined as being appropriate in only two very narrow circumstances (fraud and administrative error).

3 *Bolam* v. *Friern Hospital Management Committee* [1957] 1 WLR 582.

4 That a duty of care existed, that it was breached (i.e. the defendant's conduct fell below the legally required standard of care) and that the breach of duty caused harm to the patient. For a general discussion, see, for example, E. Jackson, *Medical Law: Text, Cases and Materials* (2nd edn) (Oxford University Press, 2010), Chapter 3, pp. 103–37.

5 *R* v. *Bateman* [1925] All ER Rep 45, 48.

6 *R* v. *Adomako* [1995] 1 AC 171, 187.

some other doctors who would testify that they might have done as she did, then it was *not open to the court* to find that defendant guilty of negligence.[7] Lord Scarman made this very clear in the House of Lords in *Maynard*, a case that involved the question of whether a judge was entitled to prefer the expert testimony of the claimants over that of the defendant:

> I have to say that a judge's 'preference' for one body of distinguished professional opinion to another also professionally distinguished is not sufficient to establish negligence in a practitioner whose actions have received the seal of approval of those whose opinions, honestly expressed, honestly held, were not preferred. If this was the real reason for the judge's finding, he erred in law ... *For in the realm of diagnosis and treatment, negligence is not established by preferring one respectable body of professional opinion to another.*[8]

The legal standard of care expected of doctors thus became both set and, essentially, policed by the medical profession itself, as the courts had denied themselves the ability to scrutinise expert evidence. Rather, they became a mere rubber stamp, whose sole role was to check for the existence of experts on the side of the defendant. It is difficult to overstate the impact of the approach in *Maynard*. The House of Lords in the case was not just unwilling to examine the medical evidence; rather, it held that a judge was *unable* to do so. This meant that the courts literally abrogated responsibility for the standard of care in medical negligence to the medical profession itself.

This would present sufficient difficulties if *Bolam* were consigned to negligence cases, but it was not. Instead it became virtually omnipresent in medical law, and was used by the courts in deciding issues as diverse – and indeed not based on technical medical skill – such as the materiality of risk in informed consent[9] and the definition of incompetent patients' 'best interests', both during and at the end of life.[10] Moreover, this use of *Bolam* was consigned only to the medical profession, and other professionals were treated much less favourably, with even their universal practices on occasion found to be negligent.[11] Thus, medical law as a subject was

[7] See M. Brazier and J. Miola, 'Bye-Bye *Bolam*: A Medical Litigation Revolution?' (2000) 8(1) *Medical Law Review*, 84. For an open affirmation of this interpretation of *Bolam* by the House of Lords, see *Maynard* v. *West Midlands Area Health Authority* [1985] 1 All ER 634.

[8] *Maynard*, 639 (emphasis added).

[9] See, for example, *Sidaway* v. *Board of Governors of Bethlem Royal Hospital* [1985] 1 All ER 643; and M. Jones, 'Informed Consent and Other Fairy Stories' (1999) 7 *Medical Law Review*, 103.

[10] See, for example, *F* v. *West Berkshire Health Authority* [1990] 2 AC 1, and *Airedale NHS Trust* v. *Bland* [1993] AC 789.

[11] See, for example, *Edward Wong Finance Co Ltd* v. *Johnson Stokes and Master* [1984] 1 AC 296.

born partly as a result of the fact that the law deferred to the perceived authority of those who practise in the rarefied environment of medicine.

This excessive deference was noted and criticised by academics such as Sheila McLean, who argued that 'the buffer which might be expected to stand between medicalisation and human rights – namely the law – has proved unwilling, unable or inefficient when asked to adjudicate on or control issues which are at best tangentially medical'.[12] However, it was also highlighted as far back as 1984 by the judiciary. In the case of *Sidaway*, the Court of Appeal criticised the use of *Bolam* in the context of informed consent for allowing doctors the freedom to essentially set their own legal standard of care:

> [The Courts] cannot stand idly by if the [medical] profession, by an excess of paternalism, denies its patients a real choice. In a word, the law will not permit the medical profession to play God ... I think that, in an appropriate case, a judge would be entitled to reject a unanimous medical view if he were satisfied that it was manifestly wrong and that the doctors must have been misdirecting themselves as to their duty in law.[13]

The higher courts were inclined to take a different view at the time, preferring to do exactly what the quote warned of, and Lord Donaldson was overruled by the House of Lords.[14] A sharp criticism of *Bolam* was made by Sally Sheldon, writing before *Bolitho*, who noted that the judges in medical law cases tended to identify more with doctors in cases and were more willing to question the decisions of patients.[15] Sheldon argued that judges' images of the parties differ wildly, and that while 'the goodwill of the altruistic medical personnel is routinely assumed, the same is not true for the claimants'.[16] The deference is therefore predicated and reliant on this identification. Using examples from the House of Lords negligence cases that considered *Bolam* in the 1980s, Sheldon demonstrated how the doctors, rather than patients, received empathy from judges, facilitating decisions based on a presumption of medical goodwill. This led to an almost inevitable prioritisation of medical privilege over patients' interests: it was almost impossible for claimants to win!

[12] S. McLean, *Old Law, New Medicine: Medical Ethics and Human Rights* (London: Pandora Publishing, 1999) p. 2.

[13] *Sidaway* v. *Board of Governors of Bethlem Royal Hospital* [1984] QB 493 (CA) 513–14.

[14] See the speech of Lord Diplock in the House of Lords in *Sidaway* v. *Bethlem Royal Hospital Governors* [1985] 1 All ER 643, and the subsequent decisions of the Court of Appeal in *Gold* v. *Haringey Health Authority* [1987] 2 All ER 888; *Blyth* v. *Bloomsbury* [1993] 4 Med. LR 151.

[15] S. Sheldon, '"A Responsible Body of Medical Men Skilled in that Particular Art ...": Rethinking the *Bolam* Test', in S. Sheldon and M. Thomson, *Feminist Perspectives on Health Care Law* (London: Cavendish, 1998).

[16] Ibid., p. 25.

While the courts were, finally, to take a different approach, and hold that judges were the final arbiters of reasonableness rather than medical professionals, this was not until 1997, in the *Bolitho* case.[17] It was then admitted by Lord Woolf – England's most senior judge – that the courts had indeed been overly deferential to doctors, but that this would now have to change.[18] Indeed, *Bolitho* is a critical case in medical law as it squarely confronts the question at the heart of *Bolam*: whether it is the courts or the medical profession that are the ultimate arbiters of the legal standard of care required of doctors. The case concerned the reasonableness or otherwise of failing to intubate a child who had suffered several respiratory episodes and whether the court was entitled to find for the claimant given that the defendant had been able to adduce evidence supporting the non-intubation. As we have seen above, under the *Maynard* interpretation of the law, the court would not have been able to make any decision other than to find for the defendant. However, in *Bolitho* the House of Lords re-established its right to make the final decision. Lord Browne-Wilkinson said that judges would be entitled to reject evidence that lacked logical force:

> the court has to be satisfied that the exponents of the body of opinion relied upon can demonstrate that such opinion has a logical basis ... if, in a rare case, it can be demonstrated that the professional opinion is not capable of withstanding logical analysis, the judge is entitled to hold that the body of opinion is not reasonable or responsible.[19]

The significance of *Bolitho*, then, lies in the fact that it reclaims for the courts the ability to judge medical evidence, rather than act as a mere rubber stamp to it. When coupled with Lord Woolf's admission that the courts had been overly deferential, the signal being sent in the civil courts was one of a change in judicial attitudes, away from an overly deferential attitude to the medical profession, towards actual scrutiny of evidence. Indeed, this view gains even more credence by the fact that the decline in deference appears to have been reflected in the criminal courts also, for the 1990s witnessed what Mason and Laurie referred to as 'a spate' of prosecutions in gross negligence manslaughter.[20] This may be seen as particularly worrying for the medical profession because negligence requires objective failure rather than conscious poor conduct. In other

[17] *Bolitho* v. *City and Hackney Health Authority* [1998] AC 232.

[18] Lord Woolf, 'Are the Courts Excessively Deferential to the Medical Profession?' (2001) 9 *Medical Law Review*, 1.

[19] *Bolitho*, 243.

[20] J. K. Mason and G. T. Laurie, *Mason and McCall Smith's Law and Medical Ethics* (8th edn) (Oxford University Press, 2011), p. 167.

words, it allows for the criminal prosecutions of doctors for grossly negligent acts or omissions that were not intended. As Mason and Laurie note, there is no requirement for an 'element of subjective wrongdoing'.[21] Moreover, cases since 2000 neglected to modify the test, which now appears to be settled law.[22] Judges, it would appear, are more willing both to question medical practice and even to use the criminal law in doing so.

The story of the law of negligence, in both the civil and criminal spheres, appears to be one of a decline in deference. In the 1990s, the civil law changed as judges granted themselves real decision-making authority, where before they had been non-interventionist. This coincided with an increase in criminal prosecutions. Of course, this should not be overplayed, as the medical profession still plays a large role in determining the legal standard of care; and since a criminal conviction will require conduct that *no* doctor would support, convictions remain rare. Nevertheless, the message from the civil courts, echoed by the public pronouncements of judges such as Lord Woolf, is clear: there *was* too much deference before, but this has now changed.

Deference in the criminal sphere

The criminal courts have traditionally shown the same deferential attitude to medical professionals. In the context of gross negligence manslaughter, the case of *Prentice* highlights the assumptions of good character made by courts.[23] In that case, two doctors made a fatal error of injecting vincristine into the spine of their patient, having been left doing a procedure they were not trained to do, unsupervised. As Ferner notes, the trial judge was at pains to point out, even when finding the defendants guilty, that they were not 'bad men [but] good men who ... were guilty of a momentary recklessness'.[24] Three famous cases, involving differing subject matter within the criminal law and decided at different times, further demonstrate the phenomenon and encapsulate the attitude of the courts. They highlight Sheldon's argument that courts feel that doctors are of good character, and also a desire to medicalise issues, thus granting a form of professional legitimacy to the defendant doctor's actions. Although not all relate to criminal prosecutions, they all concern legal issues that relate to the criminal law, such as declarations of legality. The first of these is

[21] Ibid., p. 169.

[22] See *R* v. *Mark* (Alan James) [2004] EWCA Crim 2490; *R* v. *Misra*; *R* v. *Srivastava* [2004] EWCA Crim 2375; Mason and Laurie, *Mason and McCall Smith's Law*, pp. 169–70.

[23] *R* v. *Prentice and Others* [1994] QB 302.

[24] R. E. Ferner, 'Medication Errors that Have Led to Manslaughter Charges' (2000) 321 (7270) *British Medical Journal*, 1212, 1214.

the case of *Bourne* in 1939.[25] Mr Bourne was a gynaecologist who deliberately flouted the law by performing an abortion on a fourteen-year-old who had been impregnated following a gang rape. He presented himself to the authorities soon afterwards. He was charged with unlawfully procuring a miscarriage contrary to the Offences Against the Person Act 1861. The judge, directing the jury, noted that this was not a 'normal' case of this sort, but instead one where a 'man of the highest skill' performed the procedure as 'an act of charity ... unquestioningly believing that he was doing the right thing ... *in performance of his duty as a member of a profession devoted to the alleviation of human suffering*'.[26] Mr Bourne was found not guilty by the jury. As can be seen from the italicised portion of the quote, though, not only was Mr Bourne simply assumed to be of good character, this assumption rested in large part on his being a member of the medical profession. His actions were performed with 'the highest skill', and the jury was encouraged to believe that the doctor's decision was a *professional* judgment.

In the case of *Arthur*, decided some forty years later, similar sentiments can be found.[27] Dr Arthur was originally charged with the murder of John Pearson, a baby born with Down's syndrome whom the parents did not wish to keep alive. Dr Arthur prescribed that John Pearson be given 'nursing care only', and was thus not given any food, only water and sedatives. He died some three days later. It was alleged that the lack of provision of food amounted to steps designed to bring about the death of the child. In his direction to the jury, Farquharson J also used the doctor's profession to impute a presumption of good character, and raised the spectre of professional discretion. Like Mr Bourne, he was acquitted by the jury:

> Seldom in a court could one have heard so many testaments to a man's good character, his gentleness, his skill, and his consideration for others. All features that are perhaps as remote as possible from the suggestion that he would, irresponsibly, bring about the death of a patient. The fact that he is such a man of good character does not prevent him doing such a thing ... everyone is of good character at one stage of their lives. *However, in a case of this kind, when we are talking about medical attitudes and treatment, his own career must stand him in good stead as to whether he is such a man who would do that which the prosecution submit he has done.*[28]

The third case where this attitude may also be identified is that of *Bland*. Anthony Bland was a victim of the Hillsborough disaster, and was in a persistent vegetative state. The court was asked to issue a declaratory judgment with respect to the question of whether it might be lawful to

[25] *R* v. *Bourne* [1939] 1 KB 687.
[26] Ibid., 690 (emphasis added).
[27] *R* v. *Arthur* (1981) 12 BMLR 1.
[28] Ibid. (emphasis added).

withdraw life-sustaining treatment from him. It found that the removal of treatment would be lawful. Again, the principle legal impediment was the law of murder. Lord Browne-Wilkinson actually lamented the fact that doctors could not act at the fringes of the law without having to worry about the consequences. As a vote of confidence in the beneficence of medical practitioners, it is the clearest and most strongly worded of the three in this section. The quote is long, but it merits repetition in full:

> In the past, doctors exercised their own discretion, in accordance with medical ethics, in cases such as these. To the great advantage of society, they took the responsibility of deciding whether the perpetuation of life was pointless. But there are now present amongst the medical and nursing staff of hospitals those who genuinely believe in the sanctity of human life, no matter what the quality of that life, and report doctors who take such decisions to the authorities with a view to prosecution for a criminal offence. I am not criticising such people: they are acting in accordance with their own moral standards. But their actions have made it extremely risky for a doctor to take a decision of this kind when his action may lie on the borderline of legality.[29]

Yet again, doctors are simply assumed – often merely through their belonging to the medical profession – to be of good character. The lament that doctors cannot act on the 'borderline of legality' without the law potentially becoming involved is staggering, particularly given that it was uttered by a member of the legal profession. For this reason, it can be deduced that even when the criminal law was invoked against medical professionals, they have traditionally benefitted from an extra layer of protection in judges' presumptions about them.

More recent cases in the criminal sphere

I have argued above that the civil law changed in the 1990s, and that this was accompanied by an increased willingness to use the criminal law to punish grossly negligent acts. The reason for this must be due, at least in part, to the scandals suffered by the medical profession in the 1990s. Given this, it might be expected that the same change in judicial approach might be observed as occurring through the 1990s and beyond in the criminal law as occurred in the civil law through *Bolitho*. This section examines cases that exemplify this.

The first of the cases is *Re A*.[30] This famous case involved conjoined twins, Jodie and Mary, born to devout Roman Catholic parents. Their blood was oxygenated through a common artery, and the stronger twin,

[29] *Bland*, 880.
[30] *Re A (Children) (Conjoined Twins: Surgical Separation)* [2001] Fam 147.

Jodie, was said to be keeping Mary alive. If they were not separated, doctors felt that Jodie's heart would not be able to cope with the strain and both would die within months. If they were separated, Jodie would enjoy a 'life which was worthwhile', but Mary would die in minutes.[31] The parents refused their consent to the separation on the basis of their faith. The hospital petitioned the court for a ruling regarding whether, even in the absence of the parents' consent, it would be lawful to separate the twins. The Court of Appeal noted that their decision was somewhat legalistic, but emphasised that this was deliberate:

> Much of this judgment has necessarily been rather technical, and I am conscious that some of it may seem rather remote from the deeply troubling dilemma which Jodie's and Mary's condition presents. Every member of the court has been deeply troubled by this case, but we have to decide it in accordance with the principles of existing law as we perceive them to apply to this unprecedented situation.[32]

Nevertheless, there are some aspects of interest in the case. The first is that the court did not appear to trouble itself with the question of whether the operation *should* go ahead. Rather, it confined itself to considering whether it *could*. Indeed, there is an underlying assumption that the doctors are of good conscience, though the deference of some of the judicial statements of the last section is less visible:

> Just as the parents hold firm views worthy of respect, so every instinct of the medical team has been to save life where it can be saved. Despite such a professional judgment it would, nevertheless, have been a perfectly acceptable response for the hospital to bow to the weight of the parental wish however fundamentally the medical team disagreed with it. Other medical teams may well have accepted the parents' decision. Had St Mary's done so, there could not have been the slightest criticism of them for letting nature take its course in accordance with the parents' wishes. Nor should there be any criticism of the hospital for not bowing to the parents' choice.[33]

However, there is also something of a judicial abrogation of responsibility here, as Ward LJ appeared to offer support to the treatment team irrespective of what they chose to do. While this is reasonable in these particular circumstances, it is slightly reminiscent of a *Bolam*-ite acquiescence to clinical judgement. This suspicion is further aroused when one considers that the decision of the doctors receives a ringing endorsement, while that of the parents is (albeit gently and sensitively) criticised:

[31] Ibid., 147. [32] Ibid., 258, *per* Lord Walker.
[33] Ibid., 173, *per* Ward LJ.

> There is, I sense, a lack of consistency in their approach to their daughters' welfare. In Mary's case they are overwhelmed by the legitimate, as I have found it to be, need to respect and protect her right to life ... They are fully entitled to recoil at the idea, as they see it, of killing Mary. That is wholly understandable. This lies at the core of their objection ... [Yet they] were aware of the possibility that Mary might be stillborn and they seemed reconciled to an operation which would separate Jodie from her ... In their natural repugnance at the idea of killing Mary they fail to recognise their conflicting duty to save Jodie and they seem to exculpate themselves from, or at least fail fully to face up to, the consequence of the failure to separate the twins, namely death for Jodie. In my judgment, parents who are placed on the horns of such a terrible dilemma simply have to choose the lesser of their inevitable loss. If a family at the gates of a concentration camp were told they might free one of their children but if no choice were made both would die, compassionate parents with equal love for their twins would elect to save the stronger and see the weak one destined for death pass through the gates.[34]

Needless to say, this is difficult to reconcile with the judge's statements supporting the doctors in the case, irrespective of what they decided. This is reminiscent of the criticism of *Bolam* made by Sheldon, discussed above. Moreover, as Ranaan Gillon has noticed, the parents' decision was consistent with 'widely acceptable moral reasoning', rendering the criticism less justified.[35] What can be seen in this case, then, is something of a 'deference lite', where the court will not openly defer, but still treat doctor and patient in a different manner – feeling able to criticise only one. In this particular instance, it could be because the court agreed with the doctors rather than the parents – but only one of the parties had their decision scrutinised.

What can be seen in these cases is that, despite the advent of the new approach suggested by *Bolitho* and commentators such as Lord Woolf, there is still an instinctive deference to the medical profession. In both of the cases in this section, the decisions went against the patient, and the images of doctor and patient highlighted by Sheldon remain. In the post-*Bolitho* world, there is little to suggest that doctors have reason to fear that the law will become a rampaging beast.

This can also be seen in gross negligence manslaughter cases involving medical professionals, where doctors are still treated favourably. Conviction rates are traditionally low – and certainly lower than other forms of manslaughter.[36] Moreover, even those who are found guilty can expect

[34] Ibid., 195–6.

[35] R. Gillon, 'Imposed Separation of Conjoined Twins: Moral Hubris by the English Courts' (2001) 27 *Journal of Medical Ethics*, 3, 4.

[36] See O. Quick, 'Prosecuting "Gross" Medical Negligence: Manslaughter, Discretion and the Crown Prosecution Service' (2006) 33 *Journal of Law and Society*, 421, 426–7, and

low, non-custodial sentences, and many are allowed to resume practising medicine.[37] An exception to this is Dr Feda Mulhem, who caused the death of a teenager with leukaemia by supervising the injection of a cytotoxic drug into his spine. He was sentenced to eight months' imprisonment, with an additional ten months for assaulting his wife. Nine of the total eighteen months were suspended, and since he had already spent time in jail for the assault charges he walked free from court, and thus did not spend any time in jail for the manslaughter.[38] He was subsequently disciplined by the General Medical Council (GMC), which suspended him from medical practice for twelve months.[39]

Indeed, even where the doctor concerned demonstrates a disregard for the welfare of the patient, a custodial sentence is not considered appropriate. An example of this is the case of *Misra*.[40] In this case, two senior house officers (Drs Misra and Srivastava) provided a level of post-operative care to a patient that appears to border on a disregard for his welfare. As the Court of Appeal noted:

> [the patient] was showing classic signs of infection: raised temperature and pulse rate, and lowered blood pressure. These conditions were severe, and persistent, and were or should have been obvious, if only from a glance at the relevant charts. It was equally obvious that the patient needed urgent treatment. And, notwithstanding suggestions by other members of the medical team on more than one occasion that further treatment was needed, none, or none that was appropriate, was provided. Blood results were available from the hospital computer from 9.47 on the Saturday evening. They were never obtained, nor acted upon, nor did either appellant make any enquiry whatever about the results. They did not seek help from senior colleagues. In short, infection was not diagnosed when it should have been, and not properly treated until it was far too late. The mistakes made by the appellants were elementary. Accordingly, they were negligent, and grossly so, and in consequence Sean Phillips died of toxic shock syndrome.[41]

As can be seen from the quote, suggestions from other members of the team that treatment was required were not acted on, blood test results

R. E. Ferner and S. E. McDowell, 'Doctors Charged with Manslaughter in the Course of Medical Practice, 1795–2005: A Literature Review' (2006) 99 *Journal of the Royal Society of Medicine*, 309.

37 O. Quick, 'Medical Manslaughter: The Rise (and Replacement) of a Contested Crime', in C. A. Erin and S. Ost, *The Criminal Justice System and Health Care* (Oxford University Press, 2007), pp. 29–48.

38 See C. Dyer, 'Doctor Sentenced for Manslaughter of Leukaemia Patient' (2003) 327 *British Medical Journal*, 697.

39 'In Brief' (2004) 328 *British Medical Journal*, 1218.

40 *R* v. *Misra and Another* [2004] EWCA 2375.

41 Ibid., para. 4.

were never collected, and the help of more senior colleagues was never sought. Nevertheless, the doctors involved were given suspended sentences. Moreover, given that the courts frequently comment on the good character of defendant doctors, it is noticeable that in this case the conduct of the accused is not deemed worthy of comment or condemnation. Both doctors were initially suspended by the GMC, and later cleared to return to practice, albeit initially under supervision. Dr Srivastava was cleared to practise without supervision in 2008, while Dr Misra, who had moved to India to continue to practise, was deemed not to have undertaken the further training required and was struck off the register.[42]

So what can be seen here is that, at least at first sight, the previous deference lives on, even if in a less overt form. The examples given here reflect the findings of those such as Quick and Ferner and McDowell that there are still a small number of prosecutions, and that convicted doctors almost never go to jail.

The loss of deference

But are we in the process of change? Certainly, in the recent past there have been plenty of medical 'scandals' that have raised doubts about the wisdom of assuming that all medical practitioners are of good conscience. From the surgeons at Bristol Royal Infirmary,[43] who did not recognise their own limitations (at a cost of death or brain damage to infants), to individual serial killer GP Harold Shipman, and the widespread and unlawful retention of organs by hospitals without consent,[44] the public *and the law* have seen medical professionals 'behaving badly'. As Field and Scotland note, 'high-profile medical scandals in the UK have followed one another relentlessly since the mid-1990s ... [and] shone a bright light on many dark corners of the profession and the NHS'.[45] Yet the questioning of medical power and influence – often in areas not strictly medical in nature – is not new. Rather, ethicists and others interested in medical ethics and practice have been raising such issues since the end of the Second World War, when the Nuremberg

[42] 'Sean Phillips Death Doctor Amit Misra Struck Off', BBC News, 5 May 2011 (www.bbc.co.uk/news/uk-england-hampshire-13302350).
[43] The Bristol Royal Infirmary Inquiry, *Learning From Bristol: The Report of the Public Inquiry into Children's Heart Surgery at the Bristol Royal Infirmary 1984–1995* (Cm 5207) (London: TSO, 2001).
[44] The Royal Liverpool Children's Inquiry, *Report* (London: HMSO, 2001).
[45] R. Field and A. Scotland, 'Medicine in the UK after Shipman: Has "All Changed, Changed Utterly"?' (2004) 364 *The Lancet* 40, 40.

trials of Nazi doctors demonstrated that doctors were as human as the rest of us, with a capacity for evil to match.[46] They have been proved right, and the notion of an all-powerful medical profession is now, put simply, untenable. Even the judiciary, as we have seen, has begun to change its outlook.

In the civil law, as mentioned above, the case of *Bolitho* has reinterpreted *Bolam* in a way that gives judges more control and tempers the temptation to give determinative weight to expert evidence. Similarly, the Mental Capacity Act 2005 has been the culmination of specific judicial and governmental efforts to de-*Bolam*ise the concept of best interests.[47] Informed consent is also going the same way, with English law inching ever closer towards rejecting *Bolam* outright – something that has already occurred in Canada, Australia and New Zealand.[48] The general impression from the civil law is that even judges are changing their approach – and the age of automatic deference towards the medical profession, the assumption of good character, is on the wane.

While lagging behind somewhat, the criminal law may well eventually follow suit. Already, guidelines regarding treatment at the end of life, considered sacrosanct even after *Bland*, are being scrutinised by the courts in a way that they were not even a decade ago.[49] In the case of *Burke*,[50] the first instance decision by Munby J found that GMC guidelines were contrary to the law and therefore unlawful.[51] While this was overturned on appeal, it is at least indicative of a change in judicial attitudes in the lower courts, and the upper courts will, sooner or later, follow their lead, and this may well include the criminal law. After all, judicial activism in the lower courts, such as at first instance in *Maynard*, is frequently a good indicator of the higher courts' future direction. This is not to say that there will be a glut of criminal prosecutions of medical professionals, and charges of gross negligence manslaughter

46 See J. Miola, *Medical Ethics and Medical Law: A Symbiotic Relationship* (Oxford: Hart, 2007), Chapters 1–3.

47 See J. McHale and M. Fox, *Health Care Law: Text and Materials* (2nd edn) (London: Sweet and Maxwell, 2007), Chapter 5, and M. Brazier and E. Cave, *Medicine, Patients and the Law* (4th edn) (London: Penguin, 2007) Chapter 6.

48 J. Manning, 'Informed Consent to Medical Treatment: The Common Law and the New Zealand Code of Patients' Rights' (1999) 7 *Medical Law Review*, 181; P. Skegg, 'English Medical Law and "Informed Consent": An Antipodean Assessment and Alternative' (1999) 7 *Medical Law Review*, 135; D. Chalmers and R. Schwartz, '*Rogers* v. *Whitaker* and Informed Consent in Australia: A Fair Dinkum Duty of Disclosure' (1993) 1 *Medical Law Review*, 139; and A. Maclean, 'The Doctrine of Informed Consent: Does it Exist and Has it Crossed the Atlantic?' (2004) 24(3) *Legal Studies*, 386.

49 See *Re G* [1995] 2 FCR 46.

50 *R (on the application of Burke)* v. *General Medical Council* [2004] EWCH 1879.

51 *R (Burke)* v. *General Medical Council* [2005] EWCA Civ 1003.

remain very rare. Quite rightly, there will still be an intrinsic acknowledgment of the fact that doctors almost always act with their patients' interests as their primary consideration. Moreover, as I have argued above, there are still assumptions made about doctors and their consciences in recent cases to show that deference has not entirely disappeared. However, what will almost certainly lessen eventually is the protective bubble of automatic deference that has encouraged the courts to make positive assumptions about doctors and their conduct – and grant them discretion – in criminal cases. As with the civil law, doctors will be treated more like the rest of society. The criminal law will begin to be seen as an appropriate forum for punishing errant doctors, as has begun to be the case in gross negligence manslaughter.

Conclusion: the Human Tissue Act 2004 – a blueprint for the future?

In fact, we have already seen an example of this with the Human Tissue Act 2004, relating to the removal, storage and use of human organs and tissue. This legislation arose directly as a result of the 'organ retention scandals', where it was discovered that hospitals were routinely taking and keeping organs from dead babies without the knowledge or consent of their parents.[52] Indeed, it is almost a textbook case of a decline in deference towards the medical profession amidst an increasing demand for greater accountability in the wider health care context. As Alan Milburn, then Secretary of State for Health, stated in the House of Commons, these principles were to form the bedrock of the new Act:

> *the law will be changed to enshrine the concept of informed consent.* The existing law in this area has become outdated. *The Human Tissue Act 1961 does not even contain penalties for breaches of its provisions.* The law has ill served bereaved parents in our country and causes confusion for staff. It must be changed.[53]

Amazingly, the previous legislation, the Human Tissue Act 1961, not only failed to adequately address consent issues, but did not provide for any sanction at all for non-compliance.[54] In a way that is familiar to

52 This is made clear in the explanatory notes: see Human Tissue Act 2004 Explanatory Notes, para. 5 (www.opsi.gov.uk/acts/acts2004/en/ukpgaen_20040030_en_1).

53 Speech by Alan Milburn in the House of Commons, 30 January 2001 (see BBC News, 'Milburn's Statement in Full': http://news.bbc.co.uk/1/hi/uk/1144996.stm) (emphasis added).

54 For the issues surrounding consent, see P. D. G. Skegg, *Law, Ethics and Medicine* (Oxford University Press, 1984), pp. 235–9; for the accountability issue, see I. Kennedy, 'Liability and the Human Tissue Act 1961', in I. Kennedy, *Treat Me Right: Essays on Medical Law and Ethics* (Oxford University Press, 1989).

readers of this essay, the conscience of doctors was trusted and no need for clarity was identified. The subsequent scandals demonstrated that this was not a defensible position any more, and the new Act reflects this by concentrating on autonomy and addressing the accountability issue.

The overall message, then, is somewhat mixed. On the one hand, the civil courts have deliberately sought to wrestle back control from medical practitioners, and this has filtered through into the criminal sphere with the rise in the number of gross negligence manslaughter cases since the 1980s.[55] Equally, the Human Tissue Act 2004 demonstrates that even Parliament has identified that there was too much deference in the past, and that this cannot continue. Nevertheless, it is important not to overstate the situation. As the analysis of the criminal cases highlights, it is also the case that assumptions about the conscience of doctors (both individually and as a group) still persist in courtrooms – and such assumptions are not made with respect to patients. These are the same assumptions that Sally Sheldon identified as being a part of the use of pre-*Bolitho Bolam*.

For this reason, doctors should not fear the general direction of the law. Even if they are not seen as almost immune from judicial scrutiny, as they were in the past, they remain very well regarded by judges. The former conceptualisation of the legal–medical interaction was clearly unworkable, and the current state of judicial thinking can hardly be described as unreasonable. Indeed doctors are still treated, at the very least, no worse (and arguably better) than other professionals. If over-deference has been identified and the process of eradication has begun, that might be no bad thing – for patients too deserve consideration by the law, and at times that can be seen to have been lacking.

Nevertheless, it is possible to question whether the use of the criminal law is of much benefit. As Brazier *et al.* have argued, there appears to be little deterrent effect, and it seems somewhat incongruous that non-fatal mistakes are not subject to the criminal law – no matter how serious or debilitating the injury – while fatal ones are.[56] Indeed, this state of affairs punishes the result of the mistake rather than the mistake itself, which is difficult to defend. It might be argued that gross negligence manslaughter is an inappropriate method of punishing

[55] See Quick, 'Prosecuting "Gross" Medical Negligence', 427.

[56] M. Brazier, D. Griffiths and A. Alghrani, 'A Cause for Concern?' (2010) 26 *Medical Defence Union Journal*, 11.

medical errors,[57] yet this does not necessarily conflict with arguing that the law is right to rebalance itself in favour of patients. The law is moving in the right direction in general terms, but that does not mean that the criminal law is the correct forum for addressing medical errors – particularly those involving negligence rather than intentional harm.

[57] See, for example, O. Quick, 'Medicine, Mistakes and Manslaughter' (2010) 69 *Cambridge Law Journal*, 186.

15 Criminalising medical negligence

David Archard

Introduction

Imagine that Dr Smith makes a terrible decision that has a disastrous outcome. The disastrous outcome is the death of her patient; the decision is terrible in that it displays a woeful failure on her part to respond appropriately to the particular circumstances of the patient's case. Should she be subject to the penalties of the criminal law? In England, Dr Smith might be subject to a charge of gross negligent manslaughter. Over recent years, there has been much discussion of the appropriateness of so charging doctors. In England, a number of doctors have been brought before the courts for their failures,[1] and unease has been expressed about the basis on which this has been done.[2] Criminalisation of negligent doctors has prompted two contrasting responses.[3] In broad outline, one has been welcoming of a disposition not to treat doctors as a special case and thereby to see them as being as much responsible for their failures as any other person whose terrible decisions have fatal outcomes. The other response has lamented the recourse to criminalisation and has insisted that making doctors liable in law for their mistakes is not proper, fitting or, on balance, warranted.

In what follows I shall make a case for seeing doctors as meriting special treatment. To that end I shall distinguish between good and bad reasons for seeing doctors as a 'special case', arguing that the case for criminalising the negligence of doctors must see criminal liability within

My thanks to Suzanne Ost and Margaret Brazier for very helpful comments on earlier drafts.

1 R. E. Ferner and S. E. McDowell, 'Doctors Charged with Manslaughter in the Course of Medical Practice, 1795–2005: A Literature Review' (2006) 99 *Journal of the Royal Society of Medicine*, 309.

2 O. Quick, 'Prosecuting "Gross" Medical Negligence: Manslaughter, Discretion and the Crown Prosecution Service' (2006) 33(3) *Journal of Law and Society*, 421.

3 M. Brazier and N. Allen, 'Criminalising Medical Malpractice'; O. Quick, 'Medical Manslaughter: The Rise (and Replacement) of a Contested Crime'; A. Forbes Merry, 'When Are Errors a Crime? Lessons from New Zealand', in C. A. Erin and S. Ost (eds.), *The Criminal Justice System and Health Care* (Oxford: Clarendon Press, 2007), pp. 15, 29, 67.

its more general normative context – that of the fair distribution of burdens and benefits in a criminal justice system.

I shall start by indicating briefly what matters I shall not address. Often cited in this context, their inclusion nevertheless confuses the discussion. One such matter is the more general disagreement within jurisprudence as to whether or not negligence should be a ground of criminal liability. This disagreement is often expressed as one between subjectivists, who interpret *mens rea* as requiring a guilty state of mind, a conscious disposition to undertake the act whose consequence is the harm the law seeks to prevent, and objectivists, who deny that awareness of this kind is required.[4] The debate is interesting and goes to the heart of the question of the normative foundations of the criminal law. Yet the disagreement is tangential to the current dispute. This latter is whether, if there are to be crimes whose ground of liability is negligence, they should include the acts (or omissions) of doctors with fatal outcomes. I thus do not wish to deny the propriety of prosecuting medical negligence, as some do, by appeal to a subjectivist denial of the moral appropriateness of making negligence in general a ground of liability.[5]

Similarly, I shall not be interested in the question of whether it is unfair that bad doctors who do not happen to make terrible decisions, or bad doctors whose terrible decisions are not fatal, escape punishment. This is often seen as an unfairness that derives from moral luck.[6] At stake here is an interesting question as to whether the law should only criminalise acts or should punish dispositions, states of mind and defects of character.[7] However, inasmuch as the law does attach penalties to those action-types (or omission-types) with harmful outcomes, the question presses as to whether these should include the fateful failures of doctors.

There is a further question as to whether the law should penalise non-fatal 'dangerous doctoring'. The law might do so either by adding the practice of medicine to the list of specified dangerous activities that risk serious injury or death, or by having a general endangerment

[4] For a subjectivist critique of extending liability to include negligence, see J. Hall, 'Negligent Behaviour Should be Excluded from Penal Liability' (1963) 63 *Columbia Law Review*, 632; for a statement of the contrary view, see H. L. A. Hart, 'Negligence, *Mens Rea*, and Criminal Responsibility', in H. L. A. Hart, *Punishment and Responsibility: Essays in the Philosophy of Law* (Oxford: Clarendon Press, 1968), p. 136.

[5] A. McCall Smith, 'Criminal Negligence and the Incompetent Doctor' (1993) 1 *Medical Law Review*, 336.

[6] J. C. Smith, 'The Element of Chance in Criminal Liability' [1971] *Criminal Law Review*, 63.

[7] M. Moore, 'Choice, Character and Excuse' (1990) 7 *Social Philosophy and Policy*, 29.

offence from which medical activity is not excluded.[8] My argument in this essay concerns the merits of the 'special case' that might be made on behalf of doctors. That case extends to any law that would treat doctors in the same way as others whose behaviour risks harm, whether or not a death eventuates. However, the present and immediate context of this essay is the particular law of gross negligence manslaughter, and I shall restrict myself to the kind of case this law covers. Nevertheless, it is worth acknowledging that introducing an offence of 'endangering death by dangerous doctoring' in a medical setting where the risks of possible harm are high 'could lead to an alarmingly high rate of prosecutions'.[9]

'Bad doctors'

In respect of my imagined example, I make three assumptions. First, Dr Smith is not malicious. She does not actively seek the death of her patient. Nor is she callously or cruelly indifferent to the fate of those she treats. She does not recklessly court the needless risk of her patient suffering harms. She is simply badly negligent. How is Dr Smith a *bad* doctor? Someone can be a bad doctor in the way in which someone is a bad actor or a bad footballer. They lack the skills or even the basic competence possessed by others within their chosen profession. The bad doctor, in this sense, may provide sub-optimal treatment, but he need not be a danger to his patients. There is a range of possibilities. Similarly, the bad actor may offer an underwhelming and amateurish performance, or be somebody who cannot even remember any of his lines. The bad football striker may miss more than a few goals or be the hopelessly inadequate forward barely able to kick the ball. The bad doctor in this sense is not at fault for his inadequacy, although he may be at fault for choosing to become or to continue serving as a doctor in the face of his shortcomings. Clumsiness, obtuseness, short-sightedness and stupidity may be beyond correction. Persistence in an ill-chosen profession given the presence of such non-remediable faults is not. A rigorous training and examination process should aim to exclude the hopelessly poor doctor; it need not prevent a pretty bad doctor in the indicated sense from eventually practising.

[8] For a discussion of the idea of general endangerment offences, see C. M. V. Clarkson, 'General Endangerment Offenses: The Way Forward' (2005) 32 *University of Western Australia Law Review*, 1.

[9] See O. Quick, 'Medical Killing: Need for a Specific Offence?', in C. M. V. Clarkson and S. Cunningham (eds.), *Criminal Liability for Non-Aggressive Death* (Aldershot: Ashgate, 2008), p. 168.

The bad doctor who makes the kind of terrible decisions that Dr Smith did may be bad in this first sense. But she need not be. She might indeed be capable of providing exemplary treatment. She certainly need not repeatedly make terrible decisions. She is bad in making her terrible decision inasmuch as she displays what I shall term 'lax' behaviour, marked by selfishness, indifference, laziness, inattentiveness, insensitivity, and so on. The behaviour is morally lax – unobservant, slack, slipshod, disorganised, poorly controlled – but not malicious. Nevertheless, she is blameworthy and culpable for her lax behaviour, inasmuch as defects of moral character – such as a lack of concern for the welfare of others, a literal carelessness – explain her failures. But, to repeat, culpable laxness is not maliciousness, a disposition to harm others. We may then make a presumption of generosity in respect of motivations: doctors do not want to make mistakes, nor do they wish to cause harm. Their culpability will lie in the steps they take (or do not take) to care for their patients. Their inadvertence is culpable, and their failure is not mere incompetence. To suggest as much is to conflate the two senses of 'bad'.[10]

Second, I shall assume that we can with confidence identify the relevant causal contribution of Dr Smith to her patient's death. Others may have played their role in the death; there may have been systemic failings (perhaps even culpable ones); and the patient was most likely suffering from a morbid condition that made death more likely. Nevertheless, Dr Smith culpably did or failed to do something that brought about a death that otherwise would not have resulted. For example, Dr Smith injected vincristine into the spine of a leukaemia patient instead of a vein; she did so when not properly supervised, when working long hours, and having received inadequate training in the use of this drug. Nevertheless, it was she who failed to read the warning about the possibly fatal consequences of intrathecal infusion on the drug packaging, and she who conducted the procedure.

Third, I shall assume that we can make a meaningful and practicable distinction between simple mistakes and culpable failings. Merry and McCall Smith distinguish between 'errors' and 'violations'. The former are 'unintentional failures' to plan or in doing what is needed for some goal; the latter are 'deliberate deviations' from what is normatively required to achieve a proper goal.[11] The basic and very plausible idea

[10] McCall Smith, 'Criminal Negligence'.

[11] A. F. Merry and A. McCall Smith, *Errors, Medicine and the Law* (Cambridge University Press, 2001). Errors and violations are more formally defined on pp. 74 and 101 respectively.

behind this and similar distinctions is this. All of us slip up in doing what we have set ourselves to do, and sometimes these slip-ups have bad consequences for us or for others. Nevertheless, only some of the slip-ups bespeak a failing on our part for which it may be appropriate to accuse us of wrongdoing. There is all the difference between the accusation, 'You should have looked and seen what was round the corner', and the non-judgemental observation, 'Nobody could have foreseen that coming.' We are concerned with mistakes that are not foreseen or actually intended, but which are nevertheless avoidable and for which the agent is blameworthy in not avoiding their commission.

The aggregated upshot of these three assumptions is as follows. Bad doctors, in both senses of 'bad' – namely 'incompetent' and 'lax' – nevertheless do not act maliciously with the intention of harming their patients. The distinction between the two senses of bad is complex, since a failure to take steps one might reasonably take to ameliorate the effects of an incorrigible lack of ability, an incompetence, is surely culpable ('lax') and may even be malicious. How much harm bad doctors may do is a function of both their own failings and those of the system in which they work. This latter distinction must also be handled carefully. Doctors who do not avail themselves of the systemic support that is available (accepting further training, for instance), or who persist in working in non-ideal systemic circumstances (agreeing, despite tiredness, to work extra shifts, for instance), cannot shift all the blame onto the system for what goes wrong.

A note on criminalisation

Should we criminalise Dr Smith's bad doctoring? The law is a crude instrument for securing future compliant good behaviour or for delivering just desserts in respect of errant past behaviour. It is costly in both moral and financial terms. The law must be serviced by official agencies comprising trained and continuously supervised personnel. The investigation and subsequent prosecution of anyone is a lengthy and demanding process. To be prosecuted is to be subject to a humiliating public exposure from which even the innocent suffer. Anybody who is convicted and sentenced bears a taint on their person that extends to those to whom they are close. For all these reasons – and more besides – the law should be a weapon of last resort. Most liberal jurisprudential theorists defend legal minimalism, the use of as little criminal law as is necessary to achieve ends that cannot be realised at a lesser cost.[12]

[12] A. Ashworth, *Principles of Criminal Law* (6th edn) (Oxford University Press, 2009), Chapter 2; D. Husak, *Overcriminalisation: The Limits of the Criminal Law* (New York: Oxford University Press, 2009).

Moreover, doctors who make blameworthy mistakes, but who are not prosecuted, do not escape censure. Victims or their relatives may seek financial redress for the mistakes. Of course, the compensation paid to relatives of a doctor's victim may be covered by an indemnity scheme such that the doctor personally suffers no financial penalties.[13] Errant doctors may also be reported to their professional accreditation body, such as the UK's General Medical Council (GMC). A negligent doctor may be disbarred from further medical practice. In sum, bad doctors who are not prosecuted can suffer penalties.

However, the criminal law may offer a remedy to the relatives of victims if no other satisfactory course of punitive action is open or likely to succeed. Moreover, the law has a declaratory or communicative function, independent of its role in deterring potential offenders or in giving the actual offender his or her just desserts.[14] This serves to signal to victims, their relatives and the society beyond, that there has been a public acknowledgment of wrongdoing. Any overall assessment of the value and point of criminally prosecuting bad doctors must acknowledge the extent to which non-criminal alternatives can adequately serve this role. A health commissioner or ombudsman, for example, might hear complaints and be empowered both to issue an apology and to recommend, where appropriate, further disciplinary measures against a doctor.[15]

A note on gross negligence

When should we be minded to prosecute a doctor for negligently causing death? The basic thought is that it is when, in causing someone to die, a doctor fails as a doctor and fails in so serious a fashion as to constitute her action or inaction as criminal. In the first place, the doctor owes a particular duty of care to her patients. Her negligence lies in her failure to do what is required of her *as a doctor*. Standard definitions of negligence are framed in the following general way: P is negligent if P failed to do that which it would be reasonable to expect that someone like P in the relevant respects would have done. In this manner, what P, normatively, *should* have done, but failed to do, is defined as what someone, hypothetically, *would* have done. The 'someone' in question

[13] In the UK, the National Health Service Litigation Authority (NHSLA) handles negligence claims against employees of the Service, paying out around £800 million annually in settlement of such claims.

[14] Ashworth, *Principles of Criminal Law*, Chapter 2.

[15] See, for example, the office of the New Zealand Health and Disability Commissioner (www.hdc.org.nz/complaints).

is defined relative to the class of persons of which P is a member. Thus a negligent builder fails to do what builders might be expected to do, and not what an ordinary member of the public would have done. A negligent doctor fails to discharge the duties of a doctor.

Second, to merit criminal prosecution, the negligence must be of sufficient seriousness, and 'gross' is the appropriate qualifying term. In English law, a prosecution for gross negligence manslaughter must show that the doctor's failure to discharge her duty of care to the patient caused the patient's death, and that the degree of negligence is of a different order to that which would establish a case for civil liability. The 'amount of and degree of negligence' is determinative in a criminal prosecution. To quote the judge in the case of *Bateman*, gross negligence must show 'such disregard for the life and safety of others as to amount to a crime against the state and conduct meriting punishment'.[16] Lord Mackay reiterated this view in *Adomako*, stating, the 'essence of the matter, which is supremely a jury question, is whether, having regard to the risk of death involved, the conduct of the defendant was so bad in all the circumstances as to amount in their judgment to a criminal act or omission'.[17]

The determination of sufficiency – 'such' disregard, 'so' bad – is 'supremely' a jury's. It is problematic in being circular (a crime is that which merits criminal prosecution) and in being a matter of judgement for a jury. However, there is a further problem: we are being asked to decide not just what counts as grossly negligent conduct by anyone at all, but what amounts to a negligent breach of a medical duty of care by a doctor.[18]

Bolam and special treatment

To see this problem, it is necessary first to set aside one way in which doctors' failures have been treated as special and different. The so-called '*Bolam*isation' of the law in England and Wales consisted in the specification of the standard by which doctors may be held negligent under civil law. 'A doctor is not guilty of negligence if he has acted in accordance with a practice accepted as proper by a responsible body of medical men skilled in that particular art.'[19] Despite the judge's explicit insistence on

[16] *R* v. *Bateman* [1925] 19 Cr App R 8, 11, *per* Lord Hewart CJ.
[17] *R* v. *Adomako* [1995] 2 All ER 79, 87, *per* Lord Mackay LC.
[18] See Margaret Brazier and Amel Alghrani, 'Fatal Medical Malpractice and Criminal Liability' (2009) *Professional Negligence*, 1.
[19] *Bolam* v. *Friern Hospital Management Committee* [1957] 2 All ER 118 at 122, *per* McNair J.

'responsible', the judgment served to exonerate any doctor who could show that she did what other doctors in fact might do. Indeed, the shortcomings of *Bolam* were thought to be even more serious inasmuch as the actual practice of *some* within the profession could serve as the standard. Worse still, it was the practice that some were prepared (after the event) to attest would be or might have been their own practice.[20] In short, a doctor could escape liability for negligence merely by summoning the testimony of a solitary fellow professional willing to affirm, 'For my part I would have behaved no differently.'

Bolam's treatment of doctors as a special case was deemed to be especially egregious, for the courts would seem to have allowed the culpably poor behaviour of a doctor to be exculpated by the testimony of one of their own colleagues. The legal profession's protection of a fellow professional, the doctor (perhaps in the spirit of 'brotherly solidarity' between 'ancient professions'[21]), was compounded by allowing the medical profession to define its own standards of good care. It was thus not just that the courts were protecting professionals, but that, in the case of doctors, they were prepared to do so to a degree that they would not do for non-medics.

The legal treatment of negligent doctors has changed greatly since *Bolam*. Judgments such as that in *Bolitho*[22] have affirmed the authority of courts to determine whether a standard of care is not just customary, but responsible and reasonable. Yet it is understandable that critics of *Bolam* should have expressed their dissatisfaction with the judgment by affirming that doctors are not a special case. However, it is of course possible to recognise that doctors are a special case (and to wish to see this reflected in their legal treatment) without endorsing *Bolam* or some version of it. It is consistent to affirm two propositions: first, doctors should not be held responsible for their fatal failures in exactly the same way as non-doctors should be for theirs; second, any standard of medical negligence should not be fixed by the quotidian practices of other doctors. Put another way, it might not be right to prosecute doctors for manslaughter by gross negligence, yet also not be right that the standard by which doctors could be held accountable for their negligence be that of their peers' customary practice.

Moreover, the *Bolam* judgment was informed by a reason for thinking that doctors are a special case at law that needs to be identified and

[20] A. Grubb, 'Commentary: Negligence: Causation and *Bolam*' (1998) 6 *Medical Law Review*, 378.

[21] M. Brazier and E. Cave, *Medicine, Patients and the Law* (4th edn) (London: Butterworth, 2007), p. 155.

[22] *Bolitho* v. *City and Hackney Health Authority* [1997] 4 All ER 771.

repudiated. The thought is that any court adjudicating as to medical negligence is necessarily making a medical judgement, and that such a judgement lies beyond the proper competence of a judicial institution. Non-medical opinion might in fact be a better judge as to whether medical practices are, as their own practitioners assume, responsible and reasonable.[23] Moreover, we can also distinguish between what are exclusively medical matters and what are not: questions, for instance, concerning the estimation of risk or of what counts as the exercise of proper care and attention. It is certainly clear that fundamental normative matters, such as the value of life and what should be counted as the 'best interests' of a patient, are not beyond the proper remit of a court; still less are they matters that, under an unwarranted extension of the *Bolam* test, should be adjudicated exclusively by doctors.[24]

Nevertheless, does the use of a standard of care adjudicated by courts rather than by doctors treat the latter unfairly? A doctor who knows that he will be held accountable by the standards of other doctors' behaviour knows clearly what a departure from that standard amounts to. He knows what his fellow professionals do and do not do. However, a doctor who knows that he will be held accountable by the standards that the law finds reasonable is not any less uncertain as to what he should do. Doctors are not improperly prosecuted for significant failures to conform to a non-medical standard that it is hard for them to discern and thus to follow. All of us are held to standards of care whose interpretation is a matter of judicial process. We are not, in consequence, acting in ignorance or with a substantial lack of clarity as to what we ought to do. Allowing the courts, as opposed to doctors, to set the standards against which medical professionals are judged culpable for their errors does not leave them unable to know what they must avoid doing and thus unfairly punish them if they do make errors.

Why are doctors different?

Dr Smith made a terrible decision with a fatal outcome. Should Dr Smith be treated any differently at law because she is a doctor? The view that she should not be is one Alan Merry identifies as that adopted by the New Zealand Law Reform Group: ' there should be no special deal for doctors ... it is the level of blameworthiness that should determine

[23] J. Badenoch, 'Brushes with *Bolam*: Where Will it Lead?' (2004) 72(4) *Medico-Legal Journal*, 127, 136.

[24] J. Keown, 'Doctor Knows Best?: The Rise and Rise of the *Bolam* Test' (1995) *Singapore Journal of Legal Studies*, 342. The Mental Capacity Act (2005) Section 4 offers a full statutory definition of 'best interests' (see www.legislation.gov.uk/ukpga/2005/9/section/4).

whether or not criminal prosecution follows an accidental death, not the profession of the person causing that death'.[25] It is the view endorsed by the judge in the English legal case of *Bateman*, who said: 'The law is that anybody who causes the death of anyone else – it does not only apply to a doctor, it applies to motor drivers, railway men, or signalmen, to a number of people – is criminally responsible.'[26]

The problems with this view lie ultimately in its reliance on the apparently simple, but unanalysed idea that all that should matter is whether one causes another's death with the same degree of blameworthiness. A proper analysis of this idea must take account of both the particular features of the medical profession and the background normative principles that underpin criminal liability.

Two features of the medical profession are relevant. The first is that doctors run risks that are not run by those in other jobs and professions. These risks are those of causing harm to individuals and, of course, on occasion, of causing their death. Within the medical profession not everyone runs the same risks. Neurosurgeons, anaesthetists and those in emergency care face higher risks than general practitioners. Certainly those in other jobs and professions run high risks. Lawyers risk the conviction of their clients. Those who fly planes and drive trains risk crashes. Nevertheless, it is hard to conceive of a profession whose members regularly run such high risks of harm to those they serve. Members of the armed services run the risk of being harmed and that of causing harms to others, but they do not, save in exceptional circumstances, run the risks of causing harm to civilians. Police officers similarly run risks of suffering harms. But only specialist firearms officers and the members of armed response units are exposed to the risks of causing serious harm to innocent members of the public. But they respond to exceptional and, thankfully, comparatively rare circumstances; the risks in question are not run on a daily basis.

Moreover, to a large extent, such risks are an unavoidable hazard. Lawyers can refuse to take on cases; transport workers can decline to drive their vehicles. Doctors are faced daily with a caseload of those whose conditions they must diagnose and then address. Doctors need to treat their patients; they are under constant pressure to proceed to treatment, and they must do so under conditions of complexity and uncertainty. Many doctors will know, *ex ante*, that a certain percentage of those they treat will die, and, moreover, that at least some of these deaths can be attributed in part to something they did or failed to do.

[25] Forbes Merry, 'When Are Errors a Crime?', p. 78. [26] *Bateman*, 49.

The second idea is that the work of medical professionals is of enormous value, both to the individuals treated and to the general health of society. Collectively, we need trained and proficient doctors; as individuals we owe them our wellbeing, and, on occasion, our lives. Although there are many other valuable professions, few combine the high value of their work with the attendant risks of extreme harm that are characteristic of the medical profession. Moreover, the combination is such that the value is secured only if the risks are run, and, further, that in many instances the risks and value are in direct proportion to one another. Thus, for example, a neurosurgeon is able to bring about an important and life-improving cure only if he is prepared to run the risk of a life-damaging or life-ending mistake. What seems in general to be true of the practice of medicine is that if it is done well (and even just done sufficiently well), it secures great benefits, whereas if it is done badly (and very badly especially), it brings about great harms.

Doctors necessarily take risks in the pursuit of valuable ends. This has been acknowledged in commentaries on the question of liability for medical *recklessness*.[27] However, the same acknowledgement needs to be made in respect of cases of negligence. For whether or not doctors are aware of the particular risks of some present procedure (and if their behaviour is negligent then they are not), they are generally engaging in risky behaviour that realises valuable ends. That fact is as relevant to the evaluation of their responsibility in cases of negligent unawareness of the particular risks as it is in cases of their reckless disregard for those risks.

Level of blameworthiness

The idea central to the claim that doctors are not a special case – that we may fairly punish whosoever causes a death so long as their level of blameworthiness is the same – fails to acknowledge the particular nature of the medical profession. In consequence of that failure, doctors are unfairly subjected to the risks of criminal prosecution not suffered by the members of other professions.

To illustrate the counterclaim, compare two persons, one of whom is a doctor, Smith, and the other, Jones, who works in another profession. They are equally bad, in the sense of 'lax' rather than 'incompetent'. They are, let us imagine, equally insensitive, arrogant and inattentive to

[27] Law Commission, *Working Paper on the Mental Element in Crime* (Law Commission No. 31) (London: HMSO, 1970), p. 53, cited and discussed in A. Ashworth, 'Criminal Liability in a Medical Context: The Treatment of Good Intentions', in A. Simester and A. Smith, *Harm and Culpability* (Oxford: Clarendon Press, 1996), pp. 173, 176.

others. Yet Smith's badness is more dangerous inasmuch as she is a doctor, whereas Jones's failings may occasion little damage to others. Jones's profession is such that he is unlikely ever to cause an accidental death. He is, for instance, a translator, a horticulturalist or an astronomer. Even if Jones's profession is dangerous – for example, if he is a train driver – his culpable badness in his chosen profession would need to be so much worse than Smith's to occasion the same level of harm.

A jury is asked to consider whether a doctor's negligence is 'so bad' as to merit prosecution for the causing of another's death. Yet for a non-medical professional to cause another's death, their negligence would have to be so much worse. Put another way, the question of when a doctor is 'so' negligent in causing a death (as to deserve criminal prosecution) is ambiguous between the questions of when he is 'so' negligent *as a doctor* and when he is 'so' negligent *whatever profession* he is in. Would that degree of lax inattentiveness and insensitivity have resulted in a death, regardless of his work or only inasmuch as he was a medic?

The unfairness of not treating doctors as a special case lies in not recognising that their personal failings, even if culpable, run risks of fatal harms that those same failings would not run in other professions. Moreover, such risks are not run gratuitously, but precisely in the service of valued individual and social ends that no other professions can realise to the same degree. But should not doctors be held to a higher standard of care precisely because of the risks they run? Doctors prepare for their own profession by learning what great harms they can do and what enormous benefits they might bring. They learn how to discharge their own particular and vocation-specific duty of care. This may indeed all be true. However, it still seems unreasonable to prosecute a doctor for a failing that is gross only because the standard against which it is measured is set so much higher than for all others. Lapses that would be gross in other professions would not result in the same harms; lapses in these other professions that *would* result in the same harms would be so much more serious than those characterised as 'gross' derelictions of duty by doctors.

Fairness and liability

To hold someone criminally liable for their acts is to hold that someone may be subject to legal punishment for those acts. This seems to be true as a matter of definition. Thus the justification of holding X criminally liable for his acts or omissions is the justification of punishing X for those acts or omissions. It is, in further consequence, not possible to form a final judgment as to the appropriateness or fairness of holding a doctor criminally liable for his negligent behaviour without having a background justificatory theory of punishment. What is the warrant for *punishing* Dr Smith for her terrible decision with its fatal outcome?

It lies beyond the scope of this essay to offer a complete assessment of the ethics of punishment. However, having broadly characterised the available justifications of punishment, we can show how the punishment of doctors for their gross negligence is unjustified once the particular character of the practice of medicine is acknowledged. In this manner, the case for treating doctors as a special case is further defended.

There are three broad justifications of punishment.[28] One is consequentialist and looks to the deterrent effects of punishment; a second is retributive and characterises the punishment as, in some sense, deserved, or as a fair return to the agent for her wrongdoing; whilst the third thinks of punishment as justified by its role in communicating to the agent society's view of the wrong done, and in reintegrating the wrongdoer within the community she has wronged. Leaving to one side the problems that afflict each account, and the possible ways in which each might be amplified or defended, I offer reasons why the punishment of negligent doctors is of dubious value. These reasons derive from setting the value of medical practice within the broader normative justification of punishing wrongdoing.

Grant that the principal function of punishment lies in its deterrent effects, then the justification of the practice of punishing errant doctors must lie in a broad, all-things-considered evaluation of the consequences of so doing. Note, first, that if other measures – such as professional sanctions – deter as effectively and do so at a lesser cost than criminalisation, then the latter is unjustified. This seems plausible once we acknowledge the costs of criminalisation. We have already taken note of the considerable moral and financial costs of employing the brutish weaponry of the criminal law. In respect of doctors, the crucial further point is as follows. It is not a simple matter of introducing effective deterrents that prevent a number of deaths. Were this to be the case, we could easily conclude that the benefits of the patients' lives not lost (or irredeemably damaged) as a result of punishing their otherwise negligent doctors clearly outweighs the costs incurred by using the criminal law.

[28] For good critical reviews of the literature, see R. A. Duff, 'Legal Punishment', in *The Stanford Encyclopaedia of Philosophy* (http://plato.stanford.edu/entries/legal-punishment), or R. A. Duff, *Punishment, Communication, and Community* (Oxford University Press, 2003), Chapter 1. For a good collection of writings on punishment, see R. A. Duff and David Garland (eds.), *A Reader on Punishment* (Oxford University Press, 1994).

However, the introduction of criminal sanctions may well change medical practice such that fewer lives overall are saved. This is for at least three reasons.[29] First, doctors faced with the extreme threat of prosecution will be ill-disposed to run those risks that must be run if lives are to be saved. Second, individuals faced with the threat in question may not join or will quit the profession. An understaffed health service is less well placed to save lives. Third, a culture of openness in which the mistakes and shortcomings of doctors can be discussed by their peers contributes to the melioration of medical practice. Yet fears of prosecution may inhibit such a system of honest reporting, with the loss, thereby, of the improvements with attendant benefits that would otherwise have been secured. Thus, in sum, if it is a question of an overall and all-things-considered balancing of costs and benefits, the deterrent returns of criminalising medical negligence do not trump their negative consequences.

Is punishing a doctor for her gross negligence fair? Does it in some fashion represent a proper return on her blameworthy behaviour? Once we take account of the fact that doctors run risks on our behalf, and do so to an extent that no other professional does, it is harder to see their errors, even those attributable to 'lax' characters, as meriting the same punishment as their analogues in other contexts. Non-medical professionals, who are, for instance, insensitive or careless, do not normally cause deaths. Indeed, to do so they must be so much more insensitive and careless than doctors. But equally they are not making their mistakes in the pursuit of ends of such enormous individual and social value as the practice of medicine. Remember the claim of the New Zealand Medical Law Reform Group: 'there should be no special deal for doctors . . . it is the level of blameworthiness that should determine whether or not criminal prosecution follows an accidental death'. The rebuttal of the idea that doctors merit a 'special deal' appeals to the thought that a constant quotient of blameworthiness can be found in the negligent actions of doctors and others who cause accidental death; and that it is this quotient of blameworthiness alone that determines criminal liability. This is doubly mistaken. The laxness of the culpably bad non-doctor is far less dangerous than the same dispositional failings of a doctor. Moreover, the bad doctor intends to bring about what is of great value and runs the risks he does in trying to do so, where the non-doctor can make no similar claim for her work.

[29] There is, to my knowledge, no extensive empirical research to support these suggestions. Nevertheless, there is some evidence from New Zealand. See Forbes Merry, 'When Are Errors a Crime?' Moreover, the suggestions are eminently plausible ones.

The same line of reasoning can, I think, be used to see off the claim that criminalising the bad doctor is necessary to communicate his wrongdoing and thereby reintegrate him into the moral community. Moreover, it is important once again to stress that non-criminalisation does not mean the absence of professional censure and other public forms of sanctions. Removing a bad doctor's right to practise sends out a clear message about his professional misconduct. Furthermore, the disbarred bad doctors are not thereby subject to the general social stigmatisation of being a convicted criminal that might disadvantage them from the pursuit of some other (and presumably better-suited) line of work.

Conclusion

Society needs protection from bad doctors. However, the argument developed here has given reasons not to see bad doctors who kill their patients as obviously apt for criminal prosecution. The argument has been developed against a simple claim that doctors are not a 'special case'. That claim misunderstands both the ultimate normative grounds of criminalisation and the particular nature of the medical profession. Medical practice as such is not thereby put beyond the law. Malicious doctoring should be prosecuted; health bodies may rightly be held to a standard of corporate responsibility for their systemic failures. To exempt individual doctors, in the manner argued, is not to be guilty of inconsistency in one's use of the criminal law. Rather, it is to recognise that the concept of criminal liability must be set within its proper and broader normative context, one that is sensitive to the overall justification of punishment and to the particular characteristics of that which is punished.

16 All to the good? Criminality, politics, and public health

John Coggon

Introduction

A common complaint in recent and contemporary bioethics is that public health has gone under-explored.[1] A great deal of attention has been given to questions concerning medicine and (public) health *care*, but only in the past decade has there been much interest in wider questions of public health.[2] Increasingly, texts now look beyond the normative dynamics of the practitioner–patient relationship towards broader social issues and health-related responsibilities: rather than just consider how the *un*healthy should be treated, attention rests on how people in both good health and bad should behave; what rights, obligations, and freedoms they should have; and questions of justice associated with the 'distribution' of health within a society.[3] A prominently stated

[1] The work of writers such as Dan Beauchamp belies claims that 'public health ethics' is a 'new' or unexplored area for analysis: see, e.g., D. Beauchamp, 'Public Health: Alien Ethic in a Strange Land?' (1975) 65 *American Journal of Public Health*, 1338; D. Beauchamp, 'Public Health as Social Justice' (1976) 13 *Inquiry*, 1; D. Beauchamp, 'The Neglected Tradition of Public Health' (1985) 15 *The Hastings Center Report*, 28. For a powerful critique of the dominance of a dogmatic medical ethics destructively impoverishing bioethical inquiry, see A. Dawson, 'The Future of Bioethics: Three Dogmas and a Cup of Hemlock' (2010) 24 *Bioethics*, 218.

[2] Perhaps this is best encapsulated in the shift in focus from Norman Daniels' seminal work, *Just Health Care* (Cambridge University Press, 1985), to his *Just Health* (Cambridge University Press, 2008).

[3] See, especially, D. Beauchamp and B Steinbock (eds.), *New Ethics for the Public's Health* (New York: Oxford University Press, 1999); N. Daniels, B. Kennedy, and I. Kawachi, 'Why Justice Is Good for Our Health: The Social Determinants of Health Inequalities', (1999) 128 *Daedalus*, 215; N. Kass, 'An Ethics Framework for Public Health' (2001) 91 *American Journal of Public Health*, 1776; A. Thompson, A. Robertson, and R. Upshur, 'Public Health Ethics: Towards a Research Agenda' (2003) 9 *Acta Bioethica*, 2, 157; B. Jennings, 'Frameworks for Ethics in Public Health' (2003) 9 *Acta Bioethica*, 2, 165; S. Anand, F. Peter, and A. Sen (eds.), *Public Health, Ethics, and Equity* (Oxford University Press, 2004); M. Boylan (ed.), *Public Health Policy and Ethics* (Dordrecht: Kluwer Academic Publishers, 2004); R. Bayer, L. Gostin, B. Jennings, and B. Steinbock (eds.), *Public Health Ethics: Theory, Policy, and Practice* (Oxford University Press, 2006); M. Powers and R. Faden, *Social Justice: The Moral Foundations of Public Health and Health Policy* (Oxford University Press, 2006); A. Dawson and M. Verweij (eds.), *Ethics, Prevention,*

feature of this literature is the focus on a 'population perspective' or 'population approach'.[4] In order to understand the nature of health problems, and assess what regulatory measures might be effective to resolve them, analysts are called to study health across and between populations. This permits different understandings of causation and association than can be derived when individual cases are considered alone (e.g. allowing researchers to learn about the harms of smoking tobacco, in ways that are impossible when people are only studied individually).[5] It also allows understandings of health not just as it obtains within the human organism, but to make assessments of how 'healthy' a given behaviour, practice, or lifestyle might be.

If it is accepted that the state might have some legitimate but limited concerns in people's health, a – if not *the* – purpose of scholarly inquiry in public health law and ethics is to examine and evaluate the state's 'jurisdiction' and people's own sovereignty in regard to these matters. Academic study of the scope, nature, and enforceability of health responsibilities demands scrutiny of the legitimacy and efficacy of legal, regulatory, and other measures that bear on health. Such an approach clearly chimes with Larry Gostin's influential definition of public health law as:

> the study of the legal powers and duties of the state, in collaboration with its partners ... to ensure the conditions for people to be healthy ... and of the limitations on the power of the state to constrain for the common good the autonomy, privacy, liberty, proprietary, and other legally protected interests of individuals.[6]

In my own study of public health, I have argued that the important prior question for analysts is *what makes health public*?[7] When theorists answer this question, whether they do so tacitly or explicitly, they present a commitment to a political theory, recognise its implications for concepts such as political authority and obligation, and from there work to conclusions about what issues are 'fair game' for what manner of state interference or oversight. In other words, normative works in public

and Public Health (Oxford University Press, 2007); L. Gostin, *Public Health Law: Power, Duty, Restraint* (2nd edn) (Berkeley: University of California Press, 2008); L. Gostin, *Public Health Law and Ethics: A Reader* (2nd edn) (Berkeley: University of California Press, 2010).

4 See M. Verweij and A. Dawson, 'The Meaning of "Public" in "Public Health"', in Dawson and Verweij, *Ethics, Prevention, and Public Health*, p. 13; W. Parmet, *Populations, Public Health, and the Law* (Washington, DC: Georgetown University Press, 2009).

5 G. Rose, 'Sick Individuals and Sick Populations' (1985) 14 *International Journal of Epidemiology*, 32.

6 Gostin, *Public Health Law*, p. 4.

7 J. Coggon, *What Makes Health Public?* (Cambridge University Press, 2012).

health law and ethics entail the need to explore political questions, and from there work to conclusions on the form and basis of sound regulation.

This 'methodological' point echoes that in the Nuffield Council on Bioethics' report on ethics and public health, which says:

> Public health measures raise complex questions about the relationship between the state and individuals and organisations that are affected by its policies. They also raise questions about the duties that individuals have towards each other. A substantial body of literature in political philosophy examines these relationships of duties and entitlements.[8]

Whilst I find some of the substantive ethical assertions in the report problematic,[9] I would fully endorse this statement as giving the right starting point for an analysis of the role of law and policy, and the practical scheme of rights, freedoms, and obligations that exist within society.[10] In what follows, I discuss these issues as they relate to criminal law having a place in the regulation of public health issues. By examining the term 'public health', I explore and explain how the reach of the phrase 'public health measures' is (properly) so vast that it would be difficult for anyone sensibly to postulate that criminal law could be no part of them. Then, having argued that it is not controversial to allow a role for criminal law in public health regulation, I go on to consider two issues that do give rise to complex contestation. I suggest that before recommending criminal regulation, we are well minded to consider carefully the effect of criminalisation in any instance, most particularly in regard to offences that give rise to strong moral condemnation and possible stigma, or instances where 'unhealthy' conduct is self-regarding. Equally, I urge that whilst health may form a legitimate basis of regulation, it should not be taken blithely as a sound justification.

Public health: politics in a field without boundaries

Before considering the role (if any) of criminal regulation and public health, it is necessary to consider what public health itself might mean. Although the focus of this book is on criminal law in relation to health

[8] Nuffield Council on Bioethics, *Public Health: Ethical Issues* (London: Nuffield Council on Bioethics, 2007), p. xv.

[9] See J. Coggon, 'Harmful Rights-Doing? The Perceived Problem of Liberal Paradigms and Public Health' (2008) 34 *Journal of Medical Ethics*, 798; J. Coggon, 'What Help Is a Steward? Stewardship, Political Theory, and Public Health Law and Ethics' (2011) 62(4) *Northern Ireland Legal Quarterly*, 599.

[10] For full elaboration, see Coggon, *What Makes Health Public?*, Part II. See also Coggon, 'What Help Is a Steward?', section 2.

care primarily, I would suggest (along with others) that if our focus falls on public health, it is simply not useful for (bio)ethical analyses to look only at health care. The current section will explain why this is so, and in so doing will draw out important issues involved in discourse concerning public health.

Public health means different and often contradictory things in the various contexts and ways in which it is used. Some authors recognise as much, but seek to find commonalities between definitions in order to work with a unitary conception.[11] Others assert a definition and argue that it should be employed to the exclusion of alternative understandings.[12] It seems most straightforward to me to acknowledge the differences in usage and just try to be clear in any instance what one means when referring, for example, to a 'public health measure'. Although an exhaustive account of the term's usages is not possible here,[13] a few issues are worth drawing out. First, for some, the 'public' in public health necessarily denotes governmental competence: in this sense, public health necessarily refers, at least in part, to 'public health law'.[14] The force and prominence of such perspectives underpin arguments that analysis should take place through a political lens. Consider, for example, the widely noted definition given by the United States' Institute of Medicine:

> Public health is what we, as a society, do collectively to assure the conditions in which people can be healthy. This requires that continuing and emerging threats to the health of the public be successfully countered.[15]

This definition begins descriptively: public health is something we do. But it is immediately taken to entail a normative imperative that demands political action: threats to health must be countered. It also hints towards a second understanding of public health found elsewhere in the literature. In some instances the term is taken to denote the 'collective' or 'aggregate health' of a group of people. Whilst such an understanding is conceptually tenable, it is salutary to account for the

[11] E.g. Parmet, *Populations, Public Health, and the Law*, p. 9.

[12] E.g. M. Rothstein, 'Rethinking the Meaning of Public Health' (2002) 30 *Journal of Law, Medicine and Ethics*, 144.

[13] But see Coggon, *What Makes Health Public?*, Chapter 3.

[14] See Rothstein, 'Rethinking the Meaning of Public Health'. See also Verweij and Dawson, 'The Meaning of "Public"'; Daniel Goldberg, 'In Support of a Broad Model of Public Health: Disparities, Social Epidemiology and Public Health Causation' (2009) 2 *Public Health Ethics*, 70; M. Rothstein, 'The Limits of Public Health: A Response' (2009) 2 *Public Health Ethics*, 84.

[15] Institute of Medicine, *The Future of Public Health* (Washington, DC: National Academy Press, 1988), p. 1.

following criticism taken from a paper by Richard Mohr, if we use the term public health (or population health) in this second way:

> No literal sense exists in which there could be such a thing as a public health. To say the public has a health is like saying the number seven has a color: such a thing cannot have such a property. You have health or you lack it and I have health or lack it, because we each have a body with organs that function or do not function. But the public, an aggregate of persons similarly disposed as persons, has no such body of organs with functions which work or fail.[16]

These arguments lead us squarely into a further debate: quite what does health mean, and is there too much danger, especially when we talk about *public* health, that a politicised concept of health will disguise underlying ideological norms, and itself serve to constrain people's otherwise due political freedom?[17] As a social issue, health is a normative concept,[18] and one that permits of a plurality of contradictory substantive understandings. As such, it necessarily raises (at least *prima facie*) controversy when it is advanced as the basis of policy. If the power effectively to define health is taken from people, and their externally defined health, or even health-affecting conduct, becomes (to whatever extent) a public matter, they may not accept that claims about public health benefits are purely benign.[19] I revisit this issue below, but note it here as it raises further tensions that beset bioethical analysts, who would claim to have cleanly and narrowly limited their focus by 'only' discussing public health.

Even without seeking to achieve a unitary definition of public health, it is clear that an evaluation of law and public health will entail a focus (if only in part) on state responsibilities for health and health-affecting

[16] R. Mohr, 'AIDS, Gays, and State Coercion' (1987) 1 *Bioethics*, 35, 47.

[17] See I. Illich, *Limits to Medicine – Medical Nemesis: The Expropriation of Health* (London: Marion Boyars, 1995); P. Skrabanek, *The Death of Humane Medicine and the Rise of Coercive Healthism* (Bury St Edmunds: St Edmundsbury Press, 1994); J. Metzl and A. Kirkland (eds.), *Against Health: How Health Became the New Morality* (New York and London: New York University Press, 2010).

[18] For the most notable defence of a 'naturalist' concept of health (i.e. one that claims it is purely scientific, and need not imply any imperatives), see C. Boorse, 'Health as a Theoretical Concept' (1977) 44 *Philosophy of Science*, 542; C. Boorse, 'On the Distinction between Disease and Illness' (1975) 5 *Philosophy and Public Affairs*, 49. Whilst debates continue between 'naturalists' and 'normativists' in *medical* philosophy, social policy will necessarily build on social concepts of health rather than scientific ones (even if they permit relatedness or overlap) (J. Räikkä, 'The Social Concept of Disease' (1996) 17 *Theoretical Medicine*, 353).

[19] In practice, a balance between 'internal' and 'external' perspectives needs to be reached (again, this becomes a political question, of course): see A. Sen, 'Health Achievement and Equity: External and Internal Perspectives', in Anand *et al.*, *Public Health, Ethics, and Equity*, p. 263.

behaviour. This evaluation cannot just be of health care, as we see even in the more narrow definitions of public health, such as Mark Rothstein's:

> 'Government intervention as public health' involves public officials taking appropriate measures pursuant to specific legal authority, after balancing private rights and public interests, to protect the health of the public.[20]

What Rothstein outlines may involve health care experts and expertise, but not exclusively. So given what I have outlined, I would advocate a position wherein analysts in bioethics (and beyond) who claim to be concerned about state regulation and public health be prepared to look at all manner of health-related or health-affecting issues. They cannot usefully limit themselves just to the prescribed measures explicitly legislated as 'public health powers', or simply to the practices and regulation of health care professionals.

What may be seen as a problem with my approach is that it puts public health issues in a field without boundaries. Prior to analysis, we cannot say whether it is mandatory or even permissible for the state to institute a given policy that would bear on health.[21] This leaves all potential policy that may bear on health, and all potential reasons for denying state power to interfere, on the table for a scholar looking at normative issues and public health. It is clear that this may be awkward practically. However, if we are concerned about health and measures that may or may not be instituted to advance it, it is not clear how a principled line can be drawn to exclude very broad-ranging concerns. And to emphasise, such a perspective again reinforces the claims made above that it is important to undertake an analysis of these concerns through the lens of political philosophy. If bioethics is a study of normative dynamics, bioethical analysis of 'public health measures' must be directed at the legitimacy of freedoms and obligations within a political society; with the rights and responsibilities of states, citizens, and other politically relevant actors. Of necessity, this will give rise to complex social, legal, moral, and political conflict. This is apparent from even a cursory review of the debates that are currently so prominent in the explicitly labelled 'public health ethics' literature. It requires that analysts be able to defend their normative claims by reference to political authority, accounting for competing arguments about the nature and basis of political obligation, with all the attendant moral, epistemic, and other philosophical disputes

[20] Rothstein, 'Rethinking the Meaning of Public Health', p. 146.
[21] J. Coggon, 'Public Health, Responsibility and English Law: Are There Such Things as No Smoke Without Ire or Needless Clean Needles?' (2009) 17 *Medical Law Review*, 127.

that arise. Before considering the more controversial specifics, however, I shall present what should be taken as an uncontroversial generality: regulations that bear on health may quite defensibly entail criminal measures.

Public health policy and criminal regulation

Once it is seen that 'public health measures' potentially refers to a whole range of issues far beyond a small, tightly circumscribed regulatory sphere, it becomes clear that there is a logical and defensible role for criminal intervention in issues that are deemed to relate to public health.[22] (Note that this does not preclude a legitimate role for criminal regulation, even where 'public health measures' is afforded only the most narrow of 'jurisdictions'.) If health is taken to denote physical or mental harms to the person, it is immediately apparent that many criminal wrongs, in their deterrent effect, already serve as public health measures.[23] This claim ranges from strict liability offences found in traffic regulation, designed to limit harms caused by risky driving,[24] to issues such as physical assaults prohibited under the Offences Against the Person Act 1861. Nevertheless, it is important to stress that the point I make here is a general one. At the level of principle, I am suggesting, it should be widely accepted that sometimes some imperatives relating to health will best be dealt with by the state, and most defensibly and effectively instituted using criminal regulation. Controversy arises, however, both when we seek to establish the basis of defensible regulation – a normative question – and when we try to assess the efficacy of different potential measures – an empirical question. For the former, we need an account of the relationship between morality and law. This will necessarily be controversial. For the latter, the controversy tends to arise because some things are demonstrable empirically only within a wide scheme of uncertainty and speculation. Nevertheless, the place of criminal regulation remains within the state's broader network of potential policy approaches. The array of regulatory mechanisms available to government is well illustrated in the Nuffield Council's report on public health ethics, in its presentation of the 'intervention ladder':

> The range of options available to government and policy makers can be thought of as a ladder of interventions, with progressive steps from individual freedom and responsibility towards state intervention as one moves up the ladder. In

[22] See also Fovargue in this volume.
[23] This is so even if we remain sceptical or uncertain about the precise content of physical and mental harm.
[24] See, e.g., Road Traffic Act 1988; Road Traffic Act 1991.

considering which 'rung' is appropriate for a particular public health goal, the benefits to individuals and society should be weighed against the erosion of individual freedom. Economic costs and benefits would need to be taken into account alongside health and societal benefits. The ladder of possible policy action is as follows:

> *Eliminate choice.* Regulate in such a way as to entirely eliminate choice, for example through compulsory isolation of patients with infectious disease.
>
> *Restrict choice.* Regulate in such a way as to restrict the options available to people with the aim of protecting them, for example removing unhealthy ingredients from foods, or unhealthy foods from shops or restaurants.
>
> *Guide choice through disincentives.* Fiscal and other disincentives can be put in place to influence people not to pursue certain activities, for example through taxes on cigarettes, or by discouraging the use of cars in inner cities through charging schemes or limitations of parking spaces.
>
> *Guide choices through incentives.* Regulations can be offered that guide choices by fiscal and other incentives, for example offering tax-breaks for the purchase of bicycles that are used as a means of travelling to work.
>
> *Guide choices through changing the default policy.* For example, in a restaurant, instead of providing chips as a standard side dish (with healthy options available), menus could be changed to provide a more healthy option as standard (with chips an option available).
>
> *Enable choice.* Enable individuals to change their behaviours, for example by offering participation in an NHS 'stop smoking' programme, building cycle lanes, or providing free fruit in schools.
>
> *Provide information.* Inform and educate the public, for example as part of campaigns to encourage people to walk more or eat five portions of fruit and vegetables per day.
>
> *Do nothing or simply monitor the current situation.*[25]

The intervention ladder warrants restatement here, as for the bioethicist it is important to acknowledge such an approach to regulation, rather than conceive of 'the law' as something that merely prohibits, enforces, or permits.[26] It is important too to note that, because the point is a general one, it speaks not just to advocates of the manner of liberalism that the Nuffield Council espouses in its 'stewardship model'.[27] While regulation will, in a quantitative sense, be much more limited in a libertarian state, or more pervasive in a more 'collectivist' system, in all political states there will be a role for regulations that sit on each rung of the intervention ladder.

[25] Nuffield Council on Bioethics, *Public Health: Ethical Issues*, p. 42.

[26] See, further, R. Baldwin and J. Black, 'Really Responsive Regulation' (2008) 71 *Modern Law Review*, 59; A. Ogus, 'Regulation Revisited' (2009) *Public Law*, 332.

[27] Nuffield Council on Bioethics, *Public Health: Ethical Issues*, Chapter 2.

The work for theorists, then, is to understand and explain the demands of the political theories they advocate. In so doing, they will (or should be able to) account for the legitimate function and role of the state; the scope and demands of its mandate to introduce measures that would prioritise health; and the basis for this. Thus focus should not fall on whether criminal law is *ever* appropriate, but on determining the instances when it is appropriate. As suggested above, criminal law's most natural role would seem to be as a deterrent; serving health by encouraging conduct that will cause less harm. Within a political liberal state, the function of regulation is to serve the good of the people: 'to achieve peace and security and our interests, and . . . "more commodious living"'.[28] People have different ideas about morality and the public good, and the state has a role in establishing and ensuring the conditions in which a dominant understanding defensibly prevails.[29] In the ideal, this allows a shared society with sufficient freedom for people to make their lives their own, and sufficient constraints to protect them and others from undue interferences, harms, and wrongs, and to overcome coordination problems.

As stressed above, because so many things may bear on health, it is hard to imagine a political system where criminal law has no role to play. Nevertheless, there are some matters in health policy that evoke particular wariness. Given that criminal regulation would widely be viewed as sitting near the top of the intervention ladder (though it should be noted that regulation relates to the levels of compliance, policing, and enforcement, as well as simply the letter of the law), it may be felt that there are some possible areas of public health policy where it is of special importance that criminal regulation not be permitted. In the main, it seems safe to assert that the criminalisation of matters such as driving at excessive speed or releasing noxious chemicals into the environment is not intrinsically problematic, not least as such measures can be justified by (probable) 'harm to others' type arguments.[30] If criminal regulation is the most effective means of preventing the probable harms of such things, a policy-maker can use them without having to rely on too contentious a theory of politics (or morality). However, there are two important points to raise here that relate to cases discussed in the literature where there is a perception that prejudice overpowers ethically defensible political principle. The first

[28] T. McPherson, *Political Obligation* (London: Routledge and Kegan Paul, 1967), p. 76.
[29] C. McMahon, 'Autonomy and Authority' (1987) 16 *Philosophy and Public Affairs*, 303.
[30] I do not claim that everyone happily accepts the dominance of the 'Millian paradigm' or the more faithful developments from it, but its dominance can be taken for granted as an important aspect of contemporary understandings of legitimacy in English law.

relates to (probable) harms to self that are prohibited in a regulatory mechanism supported by criminal sanction. In this instance, it is argued by some that the law has no business interfering with 'private' decisions of citizens, at least where these do not (directly) affect others. The second sort of case relates to other-regarding harms, but in circumstances where it is argued that for principled reasons it is wrong to criminalise the conduct under issue. There is not space here to provide a detailed account, but I shall exemplify each with reference to English law.

An example of contentious regulatory interference concerning self-regarding acts is found in the case of the 'Rampton smokers', who challenged England's ban on the smoking of tobacco in secure psychiatric units.[31] The applicants were detained psychiatric patients, who sought a judicial review of the Smoke-Free (Exemption and Vehicles) Regulations 2007.[32] For them and others in their situation, the detail of the Regulations essentially meant that smoking was prohibited outright. The Health Act 2006 makes provision, inter alia, 'for the prohibition of smoking in certain premises, places and vehicles' in England.[33] It is an offence under section 8 of the Act if a 'person who controls or is concerned in the management of smoke-free premises' fails to 'cause a person smoking there to stop smoking'.[34] Subject to some exceptions, detailed in the Regulations, the Act deems that enclosed or substantially enclosed public spaces and places of work are to be smoke-free. (It might be noted that, at least ostensibly, the Act is justified by reference to harms to others.) Exceptions in the Regulations include care homes, hospices, and prisons.[35] Mental health units, however, were only exempt from the ban in the Health Act until July 2008.[36] In secure units such as Rampton, because patients cannot go outside, there is no freedom to smoke. Despite a challenge to the Regulations under Article 8 – the right to respect for private and family life – and Article 14 – the right against discrimination in the enjoyment of Convention rights – of the European Convention on Human Rights (ECHR), in both the High Court and the Court of Appeal, it was found that the lack of exemption for residents at Rampton was lawful and should remain upheld. Of relevance to the current essay was the Court of Appeal's treatment of the 'protection of

31 *R (G and B)* v. *Nottinghamshire Healthcare NHS Trust; R (N)* v. *Secretary of State for Health* [2008] EWHC 1096 (Admin); *R (N)* v. *Secretary of State for Health; R (E)* v. *Nottinghamshire Healthcare NHS Trust* [2009] EWCA Civ 795.

32 The review also involved a challenge of the Trust's own non-smoking policy. This was also unsuccessful, both in the High Court and on appeal.

33 See preamble to the Health Act 2006. 34 Health Act 2006, section 8(1).

35 Smoke-Free (Exemption and Vehicles) Regulations 2007, regulation 5.

36 Ibid., regulation 10.

health' derogation to an individual's right to respect for private and family life in Article 8(2) ECHR. The Court found that:

> [A]rticle 8(2) refers to the protection of health, not the health of others. In these circumstances we accept the submission that it is a potentially legitimate aim to restrain a person's article 8 rights for the protection of [his own] health. It was necessary and proportionate for the health of both the patients, which of course includes the appellants, and others.[37]

Although this statement was made *obiter*, it is of interest that the regulatory force of the Act was held in part to be justifiable by reference to health protection even of smokers themselves. Whilst there is support for this judgment in some quarters and a desire to see the smoking ban extended, this is a clear area of controversy.[38] There is considerable scope for reasonable disagreement with the courts' decisions.[39]

Moving to other-regarding harms or imposition of risk, it is instructive to consider the controversy surrounding the criminalisation of the sexual transmission of HIV.[40] This approach to criminal regulation has given rise to criticisms from various perspectives. It may seem desirable that a criminal deterrent be instituted to stop people knowingly risking transmission of HIV. Nevertheless, this legal development has attracted condemnation on various grounds. Of two of the most salient criticisms, one is that the issue may be seen as being cast in terms of 'otherness', where there are corrosive connotations to the policy. As Jonathan Montgomery argues:

> Those being prosecuted are demonized less for their disease status than as sexual predators, and coverage shows a strong racial overtone. ... AIDS is no longer presented as a gay plague but an African one. Even more clearly the protection of victims from aggressors has emerged as the rationale for criminalization, obscuring public health issues. The innocence of the victims and the evil of the aggressor are stressed.[41]

[37] *R (N)* v. *Secretary of State for Health; R (E)* v. *Nottinghamshire Healthcare NHS Trust*, para. 71.

[38] For further analysis, see N. Allen, 'A Human Right to Smoke?' (2008) 158 *New Law Journal*, 886; Coggon, 'Public Health, Responsibility and English Law'; J. Coggon, 'No Right to Smoke in High Security Psychiatric Hospitals' (2009) 6 *Journal of Bioethical Inquiry*, 405.

[39] See the dissenting judgment of Keene LJ: *R (N)* v. *Secretary of State for Health; R (E)* v. *Nottinghamshire Healthcare NHS Trust*.

[40] *R* v. *Dica* [2004] EWCA Crim 1103; *R* v. *Konzani (Feston)* [2005] EWCA Crim 706. For more detailed discussion, see Gurnham in this volume.

[41] J. Montgomery, 'Medicalizing Crime – Criminalizing Health? The Role of Law', in C. A. Erin and S. Ost (eds.), *The Criminal Justice System and Health Care* (Oxford University Press, 2007), p. 267.

A second criticism of this development in criminal law is found in Matthew Weait's arguments concerning the proper place of responsibility for health. Weait suggests that the position fails as a sound public health measure, as it dangerously and controversially removes any responsibility from the 'victim'. In his view:

> [W]here a person is aware of the risks associated with unprotected sex and has not satisfied him- or herself that a partner is HIV negative (or free from other serious sexually transmitted infections (STIs)) the defence of consent should, in principle, be available. The reason for taking such a position is, primarily, that the transmission of HIV should be seen first and foremost as a public health issue and that everyone, not just those who are HIV positive, has a responsibility for minimising the spread of the virus.[42]

What follows from these reflections is not that criminal regulation must be seen always as something beyond a public health measure. It may be well used as part of a regulator's approach to tackling a problem. However, given both its 'moral voice' (particularly in crimes that lead to widespread social disapprobation) and its blunt dichotomising of victim and assailant, there are strong reasons to be cautious in adopting a criminal law approach. Furthermore, we might usefully read the brief points made here in light of the detailed analysis in Sara Fovargue's chapter in this volume. Examining the potential scientific, social, and regulatory contexts surrounding xenotransplantation, Fovargue demonstrates well the complex difficulties that policy-makers face. Her arguments lend support to claims that there will inevitably be a role for criminal law measures in health regulation, but also expose the intricate matrix of matters that must be accounted for before a particular measure should find favour.

Public health, social responsibility, and health as the highest law

In this final section before the conclusion, I shall briefly raise a distinct problem for bioethical analysts who are 'walking the tightrope' between an ethically attractive position and a politically defensible regulatory response. Use of the criminal law demands caution because of its potential harshness, but must be seen as a good and proper regulatory mechanism in some instances. Similarly, when given as the basis of policy (criminal or otherwise), the term 'public health' should be treated with

[42] M. Weait, 'Knowledge, Autonomy and Consent: *R.* v *Konzani*' [2005] *Criminal Law Review*, 763, 768.

caution. It must be remembered that it serves as a shorthand, and one that in practice denotes many different things. For an analyst, it is important that there be robust concepts whose meaning, import, and basis are clear. I have argued above that (bio)ethical analyses of public health are made in a political frame: when health is made public, it is deemed to be the business of the state. Of course, public health seems to present a benign idea. But precisely this presents a difficulty. If health just denoted, for example, species-typical functioning, we would have no particular reason to care about it, or (automatically) to worry about its absence. But social theorists are not discussing such a concept.[43] Ill health is a concern when it connotes concepts of harm. Though clearly conceivable as a non-normative concept, health can only ever raise the presumption that it is a concern that should affect our behaviour if it carries some normative weight. In a liberal political society, there is a whole range of competing, contradictory normative perspectives that coexist,[44] and their variation is reflected in competing conceptions of health.

Health as a social concept, then, must contain normative criteria. And these may contradict the views on or of health held by any given individual. In a liberal state that speaks to health issues, the state will necessarily have at least a partial role in defining health. How analysts respond to this will vary. It may result, for example, in an appeal to a more libertarian system,[45] a clear exposition on what health should mean and how it relates to other conditions necessary for human flourishing,[46] or appeals to a much stronger role for the state in defining health and ensuring its protection and promotion.[47] The vast spectrum of ideas about what constitutes good or acceptable health policy, and the reasoning behind them, demonstrates how critical it is that no one simply advocate a policy 'for health reasons'. In a context of complex and contestable political arguments, sound normative analysis of public health and proposals for defensible regulation require coherent background political theories, and clearly defined concepts of health. And in

[43] Some theorists (perhaps most famously, see, e.g., Norman Daniels, *Just Health*, pp. 36–46) claim to work from a naturalist conception. However, it is plain that there is only an imperative associated with health when it gives rise to normative concerns.

[44] J. Gray, *Two Faces of Liberalism* (Cambridge: Polity Press, 2000). Note, even if a reader does not accept the moral truth of Gray's position, it works from empirical observation of a clear 'practical truth'.

[45] See, especially, Skrabanek, *The Death of Humane Medicine*.

[46] See, e.g., Powers and Faden, *Social Justice*; J. Wolff and A. de-Shalit, *Disadvantage* (Oxford and New York: Oxford University Press, 2007).

[47] See, e.g., C. Bambra, D. Fox, and A. Scott-Samuel, 'Towards a Politics of Health' (2005) 20 *Health Promotion International*, 187.

a bioethical analysis, it is the political theory that allows an analyst to assess the political imperatives associated with health, given all background constraints, and other potentially competing goods. Just as it is important not to advocate criminal regulation lightly, so it is important not to leave questions begged in a position that claims to rest on the health benefits it promotes.

Conclusions

In this essay I have sought to make three key points. First, having considered the sorts of issues addressed by (bio)ethicists interested in public health, I have argued that their concerns are with the basis, manner, and scope of state responsibility for health, and, concomitantly, the responsibility assigned to other parties: the 'bioethical tightrope' that must be walked when considering public health issues is a political one. When considering what makes health public, analysis moves to issues far beyond those just relating to health care. Indeed, that the questions go beyond just health care has almost become a mantra of contemporary bioethics, and leads logically to my second point. As public health measures may come from all areas of political governance, it is not controversial to suggest that there will be instances where criminal regulation is both the most effective and the best justified approach. Notwithstanding this general proposition, however, I have expressed how cautiously someone should advocate the institution of criminal regulation as a control of health-related issues. In particular, this holds in regard to the more stigmatising crimes. Finally, I have argued that we must be careful not to accept too readily policy arguments that are baldly based on 'health benefits'. The entrenched normativity in the terms 'public' and 'health', and the clear scope for radical disagreement on their substance and import, mean that reasoned and convincing debate must be drawn from well-explicated concepts. The general idea that health policy may entail criminal measures is not of itself controversial. The complex issue for critical analysts is establishing the specific instances when it is justified.

17 Moral controversy, human rights and the common law judge

Brenda Hale
Justice of the Supreme Court of the United Kingdom

Judges are no strangers to moral controversy both inside and outside their courtrooms. Sometimes we too feel unjustly treated. In April 2010, for example, the Supreme Court upheld a declaration, made in the lower courts, that the notification requirements in the Sexual Offences Act 2003 were incompatible with the right to respect for private life, protected by Article 8 of the European Convention on Human Rights, because there was no mechanism for taking someone off the register once he was put on it indefinitely, no matter how low or non-existent was the risk that he now presented.[1] Like many issues of crime and punishment, this raised profound moral questions of repentance and redemption – is it ever morally permissible to write someone off entirely? All the judges who addressed the matter thought that it was not.

If the Court makes a declaration of incompatibility under section 4 of the Human Rights Act 1998, the offending law remains in force and anything done under it remains valid and effective. The government has three choices. First, it may use the 'fast-track' power to amend the offending Act of Parliament by a remedial order under section 10 of the 1998 Act. This receives limited parliamentary scrutiny, but is suitable for relatively straightforward corrections. Second, it may promote an Act of Parliament to put matters right. This will receive the usual full parliamentary scrutiny and is suitable for such things as creating a wholly new scheme for keeping, using or destroying DNA samples and profiles taken from people involved in the criminal justice system. Third, it can do nothing at all and wait to see whether, in due course, the European Court of Human Rights in Strasbourg agrees with the UK court. Even then, the UK is only bound in international law to put matters right. If it does nothing, as currently with the prisoners' voting rights, there is nothing that we in the courts can do about it.

[1] *R (JF)* v. *Secretary of State for the Home Department* [2010] UKSC 17, [2011] 1 AC 331.

So it was strange, to say the least, that the Prime Minister, while choosing to use the fast-track procedure to put the Sexual Offences Act right, told Parliament that the government was 'appalled' by our judgment, which he said was 'completely offensive' and 'flies in the face of common sense'.[2] It seems, therefore, that there are many intelligent and well-educated people who do not quite understand what judges do. Forgive me, therefore, for setting out a few basic principles.

First and foremost, it is a judge's job to decide the cases that come before her. She cannot take her bat in and say, 'It is all too difficult for me.' As Lord Bingham wrote, shortly before his untimely death, 'cases are brought raising novel questions, and the judges have to answer them'.[3] The French Civil Code even has a provision that the judge who refuses to judge a case shall himself be subject to punishment.[4] Sometimes we wish that we did not have to decide difficult moral questions on which we would prefer to have a more representative view. When the appellate committee of the House of Lords was asked to rule on whether artificial nutrition and hydration could lawfully be withdrawn from Anthony Bland, the Hillsborough victim left in a persistent vegetative state, they lamented the fact that Parliament had not answered such a huge moral question for them.[5] Eighteen years on, of course, Parliament has still not done so. This is either because it likes the answers given by the judges; or, more probably perhaps, because it too finds them too difficult to answer in the simple binary style of most legislation. To parliamentarians at least, many of the most controversial moral questions do not admit of a simple yes/no answer.

Second, we judges are not making it up as we go along. At the centre of the new Supreme Court of the United Kingdom is its library; the courtrooms surround it on either side and on top, but the library is at the heart of the building. In that library are the statutes, the law reports and the legal writings that go to make up the body of the law. The House of Lords did not make up the answer to the case of Anthony Bland. They found it in some well-established legal principles – in particular, the distinction between killing and letting die. Many have found this a dubious moral distinction, but so far the law and the medical profession have found it a serviceable one.

[2] Hansard, HC Debates, vol. 523, col. 955, 16 February 2011.
[3] T. Bingham, *The Rule of Law* (London: Allen Lane, 2010), p. 45.
[4] Article 4 of the *Code Civil*: 'The judge who refuses to judge, on the pretext of the silence, obscurity or insufficiency of the law, may be prosecuted as guilty of denial of justice.'
[5] *Airedale NHS Trust* v. *Bland* [1993] AC 789.

Third, however, the law does not stand still. We are used to adapting our judge-made law to meet new problems and to fit new factual situations. But the theory is that this is what the law has always been. We are simply correcting past errors or filling in the gaps that human progress has revealed. As the great Scottish Law Lord, Lord Reid, famously but ironically put it in 1971:

> Those with a taste for fairy tales seem to have thought that in some Aladdin's cave there is hidden the Common Law in all its splendour and that on a judge's appointment there descends on him knowledge of the magic words Open Sesame. Bad decisions are given when the judge has muddled the pass word and the wrong door opens. But we do not believe in fairy tales any more.[6]

But we recognise some important limits. First, we are seeking to identify and apply the existing principles of the law, extending and adapting them to meet new situations, but not turning them on their head. Second, we do not legislate. This is not so much that we defer to Parliament; it is rather a question of institutional competence. The courts can develop and adapt within existing concepts and principles, but they cannot devise whole new legislative schemes. The courts could not devise the Human Fertilisation and Embryology Act 1990 or the Human Tissue Act 2004. There are also some things that *ought* to be decided by a democratically elected Parliament rather than by the courts. Third, of course, we are mindful that changes in judge-made law operate retrospectively, so that at the very least we should stay within the bounds of what is foreseeable.[7] We recognise the importance of legal certainty, however hard that is to achieve.

Thus, for example, we cannot create new criminal offences or widen the scope of existing offences.[8] This would be to make punishable conduct that was not punishable at the time when it was carried out. In England and Wales, at least, it is also not possible to turn conduct into a criminal offence simply because the accused has done something of which the judge and the jury strongly disapprove. However much we might think it illogical, for example, that there are offences of wilful neglect and ill-treatment of certain mental patients and people who lack mental capacity, but not of adults of sound mind, or that there is an

[6] Lord Reid, 'The Judge as Lawmaker' (1972) 12 *Journal of the Society of Public Teachers of Law*, 22.

[7] A good example is the removal of the marital rape exemption by *R* v. *R (Rape: Marital Exemption)*, [1992] 1 AC 599, sanctioned by the Strasbourg Court in *SW* v. *United Kingdom* (1996) 21 EHRR 363, at [36/34], as it was 'consistent with the essence of the offence and could reasonably be foreseen'.

[8] *R* v. *Withers* [1975] AC 842; *R* v. *Rimmington* [2005] UKHL 63, [2006] 1 AC 459.

offence of causing death by gross negligence but not of causing grievous bodily harm, we cannot invent an offence to fill the gap. Having pushed at the boundaries of the common law to recognise the offence of conspiracy to corrupt public morals,[9] the House of Lords quite swiftly refused to go further and recognise an offence of conspiracy to commit a public mischief.[10]

Equally, of course, a judge cannot refuse to call criminal conduct a crime simply because she thinks that it should not be a criminal offence at all or that the accused is not morally culpable. The judge cannot throw out a case of medical manslaughter because she thinks that this is not the right way to promote the best standards of medical care. The judge cannot throw out a case of assisting suicide because she thinks that people should have the right to choose when and how to die. But here there is a well-known difference between judges and juries. Juries are free to acquit the accused if they disapprove of the law that would render his conduct punishable. There are famous examples in the field of medical manslaughter and assisted dying where this has happened. But to a lawyer, this is an arbitrary and unpredictable solution to a serious moral problem.

Back to the business of judging: as well as applying and developing the principles of the common law, judges have to interpret Acts of Parliament and other legislation. When doing this, we are in theory looking for the 'intention of Parliament'. This is mostly an illusion, because on most points that come before us, Parliament did not have any intention at all; and even if it did, we are not usually allowed to look at the parliamentary debates to find out what it was. We have to deduce the intention of the legislation from the words used, read in the light of the statutory purpose.

But it is rare for an Act of Parliament to have to be construed and applied exactly as it would have been applied when it was first passed. Statutes are said to be 'always speaking', so the words used must be made to apply to situations that would never have been contemplated when they were enacted. Thus in 2001, a 'member of the family', first used in 1920, could be held to include a same-sex partner.[11] In 1998, 'bodily harm' in a statute of 1861 could be held to include psychiatric harm.[12] And in 2011, 'violence' could be held to extend beyond physical violence into other sorts of harmful and abusive behaviour.[13]

In all of these examples, the court is seeking to further the purpose of the legislation in the moral and social world as it now is, rather than as it

[9] *Shaw* v. *DPP* [1962] AC 220. [10] *Withers*, above n. 8.
[11] *Fitzpatrick* v. *Sterling Housing Association Ltd* [2001] 1 AC 27.
[12] *R* v. *Ireland* [1998] AC 147.
[13] *Yemshaw* v. *Hounslow London Borough Council* [2011] UKSC 3, [2011] 1 WLR 433.

used to be, but to do so in a principled and predictable way that will not offend against either the intention of Parliament or the principle of legal certainty. Once again, however, it can lead us into moral controversy. Our recent decision that 'violence' covered more forms of violent conduct than physical contact was certainly not popular with some sections of the press, although it was very popular with others.

Fourth, judges can and do disagree with one another. It is our moral duty to decide the case in the way that we think legally correct. There is always scope for disagreement in the hard cases that find their way to the Supreme Court, perhaps more so in developing the common law than in interpreting statutes.[14] Some people think this is healthy, as today's dissent may point the way towards tomorrow's decision. Others may find it inconvenient, even shocking, that the law could be in such a state of doubt and uncertainty.

So the business of judging, especially at appellate level, is replete with moral choices that have to be made within an existing framework of legal principle and legislative language. But now we have a new tool, the Human Rights Act, which gives us a whole new perspective on both judge-made law and legislation. For the first time, we have to judge them both, not against the framework of existing legal principle or the language of the legislation, but against the fundamental values enshrined in the European Convention on Human Rights. And the problem with fundamental values, as Sir Stephen Sedley has pointed out, is that 'whichever one you take with you as a guide, another one is waiting round the corner with a sock full of sand'.[15] His context was the well-known French case prohibiting funfair shows where people paid a small sum to shoot dwarves from a cannon. The dwarves protested that the ban infringed their autonomy, their right to make their living as they chose from among the limited options available to them; the final court of appeal decided that such assaults on human dignity violated public morals, even when the victims were willing.[16]

This brings me to *Pretty*[17] and *Purdy*,[18] the two cases on assisting suicide in which I played a tiny part. Mrs Pretty was diagnosed as suffering from motor neurone disease in November 1999. Her condition

[14] Cf *Yemshaw* and *Jones* v. *Kaney* [2011] UKSC 13, [2011] 2 WLR 823.
[15] S. Sedley, 'Fundamental Values – But Which?', repr. in S. Sedley, *Ashes and Sparks: Essays on Law and Justice* (Cambridge University Press, 2011), p. 295.
[16] Conseil d'Etat, *Commune de Morsang-Sur-Orge* v. *Société Fun Production et M. Wackenheim*, No. 136727, 27 October 1995, AJDA 1995 Jurisprudence p. 942 (see http://lexinter.net/JPTXT2/arret_commune_de_morsang_sur_orge.htm).
[17] *R (Pretty)* v. *Director of Public Prosecutions* [2001] UKHL 61, [2002] 1 AC 800; cf [2001] EWHC Admin 788.
[18] *R (Purdy)* v. *Director of Public Prosecutions* [2009] UKHL 45, [2010] 1 AC 345.

deteriorated quickly. By late 2001, she was essentially paralysed from the neck down, had little decipherable speech and was fed by a tube. She did not have long to live, but was frightened and distressed by the suffering and indignity she would have to bear if the disease were allowed to run its natural course. She wanted her husband to provide her with the means to end her suffering when she no longer felt able to bear it, although we were always told that the final act would be hers. She asked the Director of Public Prosecutions (DPP) for an undertaking that her husband would not be prosecuted for the offence of aiding, abetting, counselling or procuring the suicide of another.[19] The DPP refused and the courts were asked to say that his refusal was in breach of her human rights. Clearly, then, there was a clash of fundamental values – respect for human life and respect for human autonomy, both ultimately grounded in the 'supervalue' of human dignity.

One problem with the *Pretty* case was that both the DPP and the courts were being asked to write a blank cheque – that Mr Pretty would not be prosecuted, no matter what the circumstances. Another problem was that, although many people would undoubtedly see this as a question of human rights, it was difficult to fit the facts into any of the rights defined in the Convention.

The right to life is protected by Article 2: 'Everyone's right to life shall be protected by law. No-one shall be deprived of his life intentionally.' This brings with it a positive obligation to protect people from risks to their lives in certain circumstances. But the English courts rejected the argument that the right to life includes the right to die and the right to choose when and how to die. The Strasbourg Court was also not convinced that the 'right to life' had a 'negative aspect' – the right not to live. Article 2 is not phrased in terms of a freedom; it is unconcerned with issues about the quality of living or what a person chooses to do with his or her life, although some of this may be protected by other articles:

> Article 2 cannot without distortion of language be interpreted as conferring the diametrically opposite right, namely a right to die; nor can it create a right to self-determination in the sense of conferring on an individual the right to choose death rather than life.[20]

So that was Article 2. Article 3 provides that 'No-one shall be subjected to torture or to inhuman or degrading treatment or punishment.' This too brings with it the right to be protected from such treatment, not only by the state, but also by others in certain circumstances. The argument

[19] Suicide Act 1961, s. 2 (as then in force).
[20] *Pretty* v. *United Kingdom*, [2002] 2 FLR 45, (2002) 35 EHRR 1, [35].

was that the right to be protected from inhuman or degrading treatment conferred the right to die with dignity. The English courts agreed that Article 3 conferred a right to human dignity, but it was a right to live with dignity, not a right to be helped to die with dignity. The Strasbourg Court pointed out that there was no complaint that the state was inflicting any ill-treatment on Mrs Pretty, or that she was not receiving adequate medical care; nor was there any ill-treatment by anyone else – far from it, as anyone who could see the devotion of Mr and Mrs Pretty to one another and the tender loving care that he provided for her during the court hearings would attest. Articles 2 and 3 had to be construed in basically the same way. They did not require the state to permit or facilitate a person's own death.

Thus far the English courts and the Strasbourg Court had reached the same conclusions and for much the same reasons. But was there another right in play? Article 8(1) provides that 'Everyone has the right to respect for his private and family life, his home and his correspondence.' Interferences can be permitted, however, if they are a proportionate means of achieving one of the legitimate aims laid down in Article 8(2). In the Divisional Court, we thought that Article 8 might well apply:

> The advantage of article 8 in this, as in many other contexts, is that it contains within it the mechanisms for balancing the various interests engaged. Article 8(1) protects the moral and physical integrity of the individual. It is possible to spell out of the right to bodily integrity the right of a competent person to refuse life-prolonging or even life-sustaining treatment, unless there is a good reason to interfere with that right under article 8(2). We are even prepared to assume, for the purpose of this argument, that it could include the right to be allowed to take one's own life, again unless there is a good reason to interfere with it under article 8(2).[21]

Somewhat to our surprise, the House of Lords disagreed: they thought that the right to respect for private and family life was a right to respect for the way one wished to live one's life, rather than for the way one wished to die. Strasbourg, however, was inclined to agree with the Divisional Court. There are several important statements of moral and legal principle in its judgment.

Although no previous case had established that the right to self-determination was contained in Article 8, the Court considered that the notion of personal autonomy was an important principle underlying the interpretation of its guarantees.[22] The extent that the state could use compulsory powers to protect people from the consequences of their

[21] [2001] EWHC Admin 788, [53]. [22] Ibid., [61].

chosen lifestyle had long been a topic of moral and jurisprudential discussion. But under the Convention case law, interference with the right to conduct one's private life as one chooses, even in physically or morally harmful ways, has to be justified under Article 8(2).[23] Imposition of medical treatment without consent would interfere with a person's physical integrity in a manner capable of engaging Article 8, even though the patient would otherwise die or suffer serious harm.[24] The way that Mrs Pretty chose to pass the closing moments of her life was part of the act of living, and she had a right to ask that this too be respected.[25] Hence, in a much-quoted passage:

> The very essence of the Convention is respect for human dignity and human freedom. Without in any way negating the principle of the sanctity of life protected under the Convention, the Court considers that it is under article 8 that notions of the quality of life take on significance. In an era of growing medical sophistication combined with longer life expectancies, many people are concerned that they should not be forced to linger on in old age or in states of advanced physical or mental decrepitude which conflict with strongly held ideas of self and personal identity.[26]

Hence the Court was 'not prepared to exclude' that preventing her by law from exercising her choice to avoid an undignified and distressing end to her life was an interference with her Article 8 rights that required justification under Article 8(2).[27]

Mrs Pretty argued that a blanket ban on all assisted suicide was disproportionate – the aim was to protect the vulnerable, but she was not vulnerable. She was a mentally competent woman who knew her own mind and was not under any pressure to make the choice. The Court accepted this, but nevertheless found that the blanket ban was not disproportionate. The more serious the harm, the more heavily weighed considerations of public health and safety in the balance against individual autonomy. Many terminally ill people would be vulnerable, and the risks of abuse were obvious. Furthermore, the government has pointed out that there was flexibility in the DPP's discretion to prosecute and in the sentences that could be imposed. Thus, in a prophetic passage:

> It does not appear to be arbitrary to the Court for the law to reflect the importance of the right to life, by prohibiting assisted suicide while providing for a system of enforcement and adjudication which allows due regard to be given in each particular case to the public interest in bringing a prosecution, as well as to the fair and proper requirements of retribution and deterrence.[28]

[23] Ibid., [62]. [24] Ibid., [63]. [25] Ibid., [64].
[26] Ibid., [65]. [27] Ibid., [67]. [28] Ibid., [76].

So Mrs Pretty lost the case both in England and in Strasbourg, and died very shortly afterwards in the distressing and undignified way that she had most feared. But she would surely have been glad that Mrs Purdy was able to take advantage of the principles that she had established in Strasbourg.

Mrs Purdy 'has led an interesting life'.[29] She has travelled widely and held many jobs, including as a music journalist, which is how she met her husband, the Cuban jazz violinist, Omar Puente. She was diagnosed with multiple sclerosis in 1995, at the age of 31. By 2001 she was permanently using a wheelchair. By 2009 she had lost the ability to carry out many basic tasks for herself and had problems with swallowing. Further deterioration was inevitable. When life became unbearable, she wanted to be able to travel abroad to a place where assisted suicide is lawful, but she would need help to do so. She asked the Director of Public Prosecutions to promulgate a policy specific to assisted suicide, which would enable her and her husband to predict the circumstances in which he could help her without risk of prosecution.

The first hurdle was to get over the House of Lords' view in *Pretty* that Article 8 was not engaged at all. Lower courts are bound by the decisions of the higher courts, even when Strasbourg has disagreed with them.[30] Contrary to popular belief, Strasbourg is not a Court of Appeal from the national courts and we do not have to do as they say, although we almost always do. So the lower courts had not had much option but to dismiss the claim. The House of Lords, however, did not have much difficulty in accepting that, despite its guarded language, the *Pretty* case had established that Article 8 was engaged and that any interference with the way in which Mrs Purdy chose to pass the closing moments of her life had to be justified.

Once over this hurdle, the argument was that to be justified, the interference had to be 'in accordance with the law'. The blanket ban was not enough, because the very flexibility commended by Strasbourg meant that it was not sufficiently foreseeable when a person might or might not be prosecuted or punished for his assistance. The Convention principle of legality requires that the law be accessible and precise enough for a person to be able to regulate their conduct accordingly. In *Pretty*, the only argument had been the proportionality of the blanket ban. Mrs Purdy relied on the passage quoted above to show that the interference would not always be proportionate and therefore that

[29] N. Cartwright, '40 Years On: Is the Suicide Act Fit for Purpose?' (2009) 17 *Medical Law Review* 467.

[30] *Lambeth London Borough Council* v. *Kay* [2006] UKHL 10, [2006] 2 AC 465.

greater clarity was required. The House of Lords unanimously agreed with Mrs Purdy and the unfortunate DPP had to go away and devise some more precise guidelines.[31] Inevitably, in doing so, he would not be able to please everybody. Of course, many would say that he should not have to do so – that Parliament is abdicating its responsibility by failing to legislate for a specific scheme.[32]

Broadly speaking, the factors tending in favour of prosecution fall into three categories. First are those casting doubt on the autonomy of the so-called 'victim' – being under the age of 18, lacking the capacity to decide, not having reached a voluntary, clear, settled and informed decision to commit suicide, and not having communicated this clearly and unequivocally to the so-called 'suspect'. Second are those casting doubt on the suspect's motivation – a history of violence or abuse against the victim, putting pressure on the victim, not ensuring that no one else has done so, helping someone who could have done it for themselves, and not being wholly motivated by compassion – for example, standing to gain in some way from the death. Third are those aimed against stranger helpers – those who encourage or assist by providing information in books, pamphlets or over the Internet, who help multiple victims who are unknown to one another, who are paid to help, who are involved in providing premises where people can commit suicide, and, above all, who are health care professionals, professional carers or people in authority over victims in their care. Medically assisted suicide this is not, although two doctors who have helped patients to go to Switzerland have not in fact been prosecuted.[33]

Noticeably absent from the list of factors either for or against prosecution is any response to my plea in *Purdy*:

> It is not for society to tell people what to value about their own lives. But it may be justifiable for society to insist that we value their lives even if they do not. In considering the factors for and against prosecution in the Daniel James case [the young man with serious spinal injuries sustained playing rugby but not terminally ill] the DPP did not focus upon the reasons why Daniel wished to die ... I would hope that some attention would be paid to the reasons why the person (whose convention rights are engaged) wished to be helped to end his or her life ... If it is the convention which is leading us to ask the Director for greater clarity, a relevant question must be in what circumstances the law is justified in interfering with a genuinely autonomous choice.[34]

[31] Crown Prosecution Service, *Policy for Prosecutors in Respect of Cases of Encouraging or Assisting Suicide* (London: CPS, February 2010).

[32] Cartwright, '40 Years On'.

[33] See, e.g., '"Dr Death" Michael Irwin Has Helped at Least Nine People End their Lives', *The Telegraph*, 26 June 2010.

[34] *Purdy*, [68].

In other words, is the situation in which the 'victim' finds herself, the reason why she wants to die, relevant to the culpability of the person who helps her to do so? Most discussions of this subject in the media assume a terminally ill patient, such as Mrs Pretty, who wishes to die slightly earlier than nature would allow. These were the circumstances in which Lord Falconer's amendment to the Coroners and Justice Bill in 2009 would have made it lawful to help a person to travel abroad to a country where assisted dying is lawful.[35] These are the cases where public support for assisted dying is the greatest. Lord Falconer's amendment would not have helped Daniel James, who was paralysed but not terminally ill. The Strasbourg Court in *Pretty* was concerned about a rather wider group of people, those who did not want 'to linger on in old age or states of advanced physical or mental decrepitude' and should not be forced to do so.[36] But we may well ask what right we have to force anyone who finds their life intolerable, for whatever reason, to linger on when they do not wish to do so. When does anyone have a duty to stay alive? The law does not expect this of anyone – even of a mentally competent pregnant woman.[37] So why should people who, for whatever reason, lack the means to take their own lives be denied help to do so?

The DPP's interim guidelines[38] did refer to the illness or disability of the victim, but these have gone from the final version. No one would want to suggest that people who are terminally ill, still less that people with serious disabilities, are any less valuable than anyone else, or that it is any less worthwhile keeping them alive for as long as they want to live. It must be a great relief to those who occupy the 'mobile benches' in the House of Lords (as wheelchair-using peers describe themselves) to have got this idea out of the way. The Dutch law depends on unbearable suffering that stems from a medical condition with no prospect of improvement.[39] The Swiss law has no medical criteria.[40] It leads us inevitably back to autonomy and disinterest as the governing criteria, rather than the situation in which the 'victim' finds herself. I think that this is the right approach, but it is unlikely to be one that Parliament would ever endorse.

[35] Hansard, HL Debates, vol. 712, col. 597, 7 July 2009.
[36] *Pretty* v. *United Kingdom*, [65].
[37] *St George's Healthcare NHS Trust* v. *S* [1999] Fam 26.
[38] Crown Prosecution Service, *Interim Policy for Prosecutors in Respect of Cases of Assisted Suicide* (London: CPS, 2009).
[39] Termination of Life on Request and Assisted Suicide (Review Procedures) Act 2002.
[40] See S. A. Hurst and A. Mauron, 'Assisted Suicide and Euthanasia in Switzerland: Allowing a Role for Non-Physicians' (2003) 326 *British Medical Journal*, 271.

I mention these cases, not because they are directly relevant to the criminal liability of doctors and scientists for what they do. The DPP's guidelines on assisting suicide will not make it any more likely that a doctor who does provides medical help to die will escape prosecution for assisting suicide in future, still less that he will do so for assisting a patient to die in rather more direct ways. I mention them to illustrate the added moral and political dimensions that human rights have brought to the business of judging. There are several of these new dimensions.

First, we now have to approach both judge-made and statute law through the lens of human rights – principally the European Convention, but also the other international human rights conventions to which we are party. Although most of these other conventions have not been made part of our domestic law, they are frequently invoked by Strasbourg in support of what it calls the 'evolutive interpretation' of the European Convention rights. Since the 1998 Act came into force, people in the UK have had the right to expect that their law conforms to certain standards, and not just to the standards adopted from time to time by judges and politicians. It has expanded the role of the courts in at least three ways: ruling on the compatibility of Acts of Parliament with the Convention, determining whether a public authority has acted compatibly with the Convention rights, and developing the common law so that the courts respect human rights in adjudicating between individuals. To quote Sir Stephen Sedley again, human rights 'are by nature political, for they seek to condition how states treat individuals'.[41] By requiring the judges to adjudicate on them, Parliament has increased their small 'p' political role.

Second, pervading most of the Convention's guarantees is the search for a fair balance, between the rights of the individual and the interests of the community as a whole. This does not apply to the absolute rights, in particular to the right not to be subjected to torture or to inhuman or degrading treatment: this is not allowed, no matter how much good it would do for the community as a whole – for example, send a dangerous foreign terrorist back to be tortured by his own people or to ill-treat a suspect to get him to reveal where the explosives have been laid. But in most cases there is a balance to be struck between the freedom of the individual to live her life as she chooses and the interests of the community in restricting that freedom.

[41] S. Sedley, 'Human Rights: A Twenty-First Century Agenda', in Sedley, *Ashes and Sparks*, p. 348.

In effect, therefore, the essence of the Convention is compromise – there are a few moral absolutes, but not many, and even those that appear absolute are not quite what they seem. For example, the meaning to be attached to 'inhuman or degrading treatment' under Article 3 undoubtedly changes over time. We have had to look at the conditions in prisons and psychiatric hospitals and units with this in mind,[42] although it is fair to say that Strasbourg will forgive a great deal under the heading of 'medical necessity'.[43] They seem to think that to be subjected to what the doctors genuinely (and defensibly) think best cannot be inhuman or degrading. So our answer to the question, 'Is compromise ethical?', would have to be that, in law, it is not only ethical but frequently required.

Third, therefore, it is a myth to think that human rights standards are universal either in time or place. The European Convention, along with the Universal Declaration of Human Rights and its children, the International Covenants on Civil and Political Rights and on Economic, Social and Cultural Rights, have their roots in the 'possessive individualism' of the Enlightenment. But the European Convention has long been recognised to be a 'living instrument', which can grow and develop with the times. I must say that I prefer Lord Sankey's analogy of a 'living tree', which can grow and develop, but only within its natural limits.[44] This at least suggests that there are some natural limits – that the tree is not a beanstalk.

The trouble for us is knowing what those natural limits are. One feature of Strasbourg's evolutive approach to the interpretation of the Convention has been the narrowing of the margin of appreciation accorded to member states, particularly when assessing the proportionality of interfering with rights – in other words, in striking the balance between the individual and the community. The Strasbourg Court appeals to a strong consensus among the member states, but how is this so-called consensus discovered? Sometimes it is judged by the standards to be found in the domestic legislation of member states;[45] sometimes by evolving attitudes and beliefs among the member states;[46] and sometimes it seems to be some way ahead of both.[47] The Court decides by reference to a consensus among member states that a particular interference cannot be justified. But it takes a variable and sometimes

[42] *R (Munjaz)* v. *Mersey Care NHS Trust* [2005] UKHL 58, [2006] 2 AC 148.
[43] *Herczegfalvy* v. *Austria* (1993) 15 EHRR 437.
[44] *Edwards* v. *Governor General of Canada* [1930] 2 AC 124, 136.
[45] *S and Marper* v. *United Kingdom* (2009) 48 EHRR 50.
[46] *Marckx* v. *Belgium* (1979–80) 2 EHRR 330.
[47] *Hirst* v. *United Kingdom* (2006) 42 EHRR 41.

unscientific approach to discovering this consensus. This does not make life any easier for domestic courts that are doing their best to 'keep pace with the Strasbourg jurisprudence as it develops over time: no more, but certainly no less'.[48]

The danger, of course, is that the more the Strasbourg Court is seen by politicians to be stepping into an arena that is properly theirs, the more the politicians will react by attacking the decisions of that Court and of the domestic courts that follow them. We have seen this recently in relation to the prisoners' voting issue. In *Hirst* v. *United Kingdom*, Strasbourg decided that a blanket ban on prisoners' voting could not be justified, leaving it to Parliament to decide what to do about it. Parliament proved remarkably reluctant to do anything, so, in December 2010, Strasbourg gave them six months. At the time of writing, nothing has been done. The Prime Minister has said that the whole idea makes him feel physically sick. The Master of the Rolls has recently commented:

> We may think that it is inappropriate that Strasbourg pokes its nose into the votes for prisoners issue on the basis that it should be left to our Parliament to decide. However, if Strasbourg said that votes for criminals was a matter for national legislatures, it may be that a dictator might see this as a green light to depriving his enemies of the vote by trumping up charges to bring against them. It may be thought to be a small price to pay for a civilised Europe that we sometimes have to adapt our laws a little.[49]

The Convention is meant to protect the interests of the individual, and particularly the unpopular individual, against the will of the majority. Women did in the end succeed in persuading men to give them the vote. But they should have got it because it was their fundamental right, not because the men chose to give it to them.

This all goes to show that the business of judging is now very different from the business of judging in the mid-twentieth century, when I first thought of learning the law. We are drawn into moral and political controversy to a much greater extent than we used to be. But for the most part this is because of the challenges thrown up by the cases that come to us, and by the increasing variety of the tools that we have been handed to solve them, and not because we have gone out looking for trouble.

[48] *R (Ullah)* v. *Special Adjudicator* [2004] UKHL 26, [2004] 2 AC 323, para. 20.

[49] Lord Neuberger of Abbotsbury, 'Who Are the Masters Now?', Second Lord Alexander of Weedon Lecture, 6 April 2011, [64] (see www.judiciary.gov.uk/NR/rdonlyres/F402769A-196C-47E3-B5D9-AAA1EEE096F3/0/mrspeechweedonlecture110406.pdf).

Index